From NEWTON *to* EINSTEIN: Changing Conceptions of THE UNIVERSE

BY
BENJAMIN HARROW, PH.D.
SECOND EDITION, REVISED AND ENLARGED
WITH ARTICLES BY PROF. EINSTEIN, PROF. J. S. AMES (JOHNS HOPKINS), SIR FRANK DYSON (ASTRONOMER ROYAL), PROF. A. S. EDDINGTON (CAMBRIDGE) AND SIR J. J. THOMSON (PRESIDENT OF THE ROYAL SOCIETY)
Portraits and Illustrations

FOR MORE SCIENCE CLASSICS:

https://www.bit.ly/shouldersofgeniuses

First Edition, May, 1920
Second Edition, October, 1920

PREFACE

Einstein's contributions to our ideas of time and space, and to our knowledge of the universe in general, are of so momentous a nature, that they easily take their place among the two or three greatest achievements of the twentieth century. This little book attempts to give, in popular form, an account of this work. As, however, Einstein's work is so largely dependent upon the work of Newton and Newton's successors, the first two chapters are devoted to the latter.

    B. H.

PREFACE TO SECOND EDITION

The preparation of this new edition has made it possible to correct errors, to further amplify certain portions of the text and to enlarge the ever-increasing bibliography on the subject. Photographs of Professors J. J. Thomson, Michelson, Minkowski and Lorentz are also new features in this edition.

The explanatory notes and articles in the Appendix will, I believe, present no difficulties to readers who have mastered the contents of the book. They are in fact "popular expositions" of various phases of the Einstein theory; but experience has shown that even "popular expositions" of the theory need further "popular introductions."

I wish to take this opportunity of thanking Prof. Einstein, Prof. A. A. Michelson of the University of Chicago, Prof. J. S. Ames of Johns Hopkins University, and Professor G. B. Pegram of Columbia University for help in various ways which they were good enough to extend to me. Prof. J. S. Ames and the editor of *Science* have been kind enough to allow me to reprint the former's excellent presidential address on Einstein's theory, delivered before the members of the American Physical Society.

CONTENTS

PREFACE iii

PREFACE TO SECOND EDITION iv

NEWTON 1

THE ETHER AND ITS CONSEQUENCES 19

EINSTEIN 30

    References 54

APPENDIX 58

    NOTE 1 (page 21) 58

    NOTE 2 (page 27) 58

    NOTE 3 (page 30) 59

    NOTE 4 (page 51) 60

    NOTE 5 (page 55) 63

NOTE 6 (page 56) 63

NOTE 7 (page 57) 64

NOTE 8 (page 59) 64

NOTE 9 (page 67) 65

TIME, SPACE, AND GRAVITATION1 68

EINSTEIN'S LAW OF GRAVITATION1 74

THE DEFLECTION OF LIGHT BY GRAVITATION AND THE EINSTEIN THEORY OF RELATIVITY.1 94

   SIR FRANK DYSON the Astronomer Royal 94

   Professor A. S. Eddington ROYAL OBSERVATORY 97

   Sir J. J. Thomson President of the Royal Society 99

NEWTON

"Newton was the greatest genius that ever existed."—*Lagrange, one of the greatest of French mathematicians.*

"The efforts of the great philosopher were always superhuman; the questions which he did not solve were incapable of solution in his time."—*Arago, famous French astronomer.*

EINSTEIN

"This is the most important result obtained in connection with the theory of gravitation since Newton's day. Einstein's reasoning is the result of one of the highest achievements of human thought."—*Sir J. J. Thomson, president of the British Royal Society and professor of physics at the University of Cambridge.*

"It surpasses in boldness everything previously suggested in speculative natural philosophy and even in the philosophical theories of knowledge. The revolution introduced into the physical conceptions of the world is only to be compared in extent and depth with that brought about by the introduction of the Copernican system of the universe."—*Prof. Max Planck, professor of physics at the University of Berlin and winner of the Nobel Prize.*

SIR ISAAC NEWTON

I

NEWTON

In speaking of Newton we are tempted to paraphrase a line from the Scriptures: Before Newton the Solar System was without form, and void; then Newton came and there was light. To have discovered a law not only applicable to matter on this earth, but to the planets and sun and stars beyond, is a triumph which places Newton among the super-men.

What Newton's law of gravitation must have meant to the people of his day can be pictured only if we conceive what the effect upon us would be if someone—say Marconi—were actually to succeed in getting into touch with beings on another planet. Newton's law increased confidence in the universality of earthly laws; and it strengthened belief in the cosmos as a law-abiding mechanism.

Newton's Law. The attraction between any two bodies is proportional to their masses and inversely proportional to the square of the distance that separates them. This is the concentrated form of Newton's law. If we apply this law to two such bodies as the sun and the earth, we can state that the sun attracts the earth, and the earth, the sun. Furthermore, this attractive power will depend upon the distance between these two bodies. Newton showed that if the distance between the sun and the earth were doubled the attractive power would be reduced not to one-half, but to one-fourth; if trebled, the attractive power would be reduced to one-ninth. If, on the other hand, the distance were halved, the attractive power would be not merely twice, but four times as great. And what is true of the sun and the earth is true of every body in the firmament, and, as Professor Rutherford has recently shown, even of the bodies which make up the solar system of the almost infinitesimal atom.

This mysterious attractive power that one body possesses for another is called "gravitation," and the law which regulates the motion of bodies when under the spell of gravitation is the law of gravitation. This law we owe to Newton's genius.

Newton's Predecessors. We can best appreciate Newton's momentous contribution to astronomy by casting a rapid glance over the state of the science prior to the seventeenth century—that is, prior to Newton's day. Ptolemy's conception of the earth as the center of the universe held undisputed sway throughout the middle ages. In those days Ptolemy was in astronomy what Aristotle was in all other knowledge: they were the gods who could not but be right. Did not Aristotle say that earth, air, fire and water constituted the four elements? Did not Ptolemy say that the earth was the center around which the sun revolved? Why, then, question further? Questioning was a sacrilege.

Copernicus (1473–1543), however, did question. He studied much and thought much. He devoted his whole life to the investigation of the movements of the heavenly bodies. And he came to the conclusion that Ptolemy and his followers in succeeding ages had expounded views which were diametrically opposed to the truth. The sun, said Copernicus, did not move at all, but the earth did; and far from the earth being the center of the universe, it was but one of several planets revolving around the sun.

The influence of the church, coupled with man's inclination to exalt his own importance, strongly tended against the acceptance of such heterodox views. Among the many hostile critics of the Copernican system, Tycho Brahe (1546–1601) stands out pre-eminently. This conscientious observer bitterly assailed Copernicus for his suggestion that the earth moved, and developed a scheme of his own which postulated that the planets revolved around the sun, and planets and sun in turn revolved around the earth.

The majority applauded Tycho; a small, very small group of insurgents had faith in Copernicus. The illustrious Galileo (1564–1642) belonged to the minority. The telescope of his invention unfolded a view of the universe which belied the assertions of the many, and strengthened his belief in the Copernican theory. "It (the Copernican theory) explains to me the cause of many phenomena which under the generally accepted theory are quite unintelligible. I have collected arguments for refuting the latter, but I do not venture to bring them to publication." So wrote Galileo to his friend, Kepler. "I do not venture to bring them to publication." How significant of the times—of any time, one ventures to add.

Galileo did overcome his hesitancy and published his views. They aroused a storm. "Look through my telescope," he pleaded. But the professors would not; neither would the body of Inquisitors. The Inquisition condemned him: "The proposition that the sun is in the center of the earth and immovable from its place is absurd, philosophically false and formally heretical; because it is expressly contrary to the Holy Scriptures." And poor Galileo was made to utter words which were as far removed from his thoughts as his oppressors' ideas were from the truth: "I abjure, curse and detest the said errors and heresies."

The truth will out. Others arose who defied the majority and the powerful Inquisition. Most prominent of all of these was Galileo's friend, Kepler. Though a student of Tycho, Kepler did not hesitate to espouse the Copernican system; but his adoption of it did not mean unqualified approval. Kepler's criticism was particularly directed against the Copernican theory that the planets revolve in circles. This was boldness in the extreme. Ever since Aristotle's discourse on the circle as a perfect figure, it was taken for granted that motion in space was circular. Nature is perfect; the circle is perfect; hence, if the sun revolves, it revolves in circles. So strongly were men imbued with this "perfection," that Copernicus himself fell victim. The sun no longer moved, but the earth and the planets did, and they moved in a circle. Radical as Copernicus was, a few atoms of conservatism remained with him still.

Not so Kepler. Tycho had taught him the importance of careful observation,—to such good effect, that Kepler came to the conclusion that the revolution of the earth around the sun takes the form of an ellipse rather than a circle, the sun being stationed at one of the foci of the ellipse.

To picture this ellipse, we shall ask the reader to stick two pins a short distance apart into a piece of cardboard, and to place over the pins a loop of string. With the point of a pencil draw the loop taut. As the pencil moves around the two pins the curve so produced will be an ellipse. The positions of the two pins represent the two foci.

Kepler's observation of the elliptical rotation of the planets was the first of three laws, quantitatively expressed, which paved the way for Newton's law. Why did the planets move in just this way? Kepler tried to answer this also, but failed. It remained for Newton to supply the answer to this question.

Newton's Law of Gravitation. The Great Plague of 1666 drove Newton from Cambridge to his home in Lincolnshire. There, according to the celebrated legend, the philosopher sitting in his little garden one fine afternoon, fell into a deep reverie. This was interrupted by the fall of an apple, and the thinker turned his attention to the apple and its fall.

It must not be supposed that Newton "discovered" gravity. Apples had been seen to fall before Newton's time, and the reason for their return to earth was correctly attributed to this mysterious force of attraction possessed by the earth, to which the name "gravity" had been given. Newton's great triumph consisted in showing that this "gravity," which was supposed to be a peculiar property residing in the earth, was a universal property of matter; that it applied to the moon and the sun as well as to the earth; that, in fact, the motions of the moon and the planets could be explained on the basis of gravitation. But his supreme triumph was to give, in one sublime generalization, quantitative expression to the motion regulating heavenly bodies.

Let us follow Newton in his train of thought. An apple falls from a tree 50 yards high. It would fall from a tree 500 yards high. It would fall from the highest mountain top several miles above sea level. It would probably fall from a height much above the mountain top. Why not? Probably the further up you go the less does the earth attract the apple, but at what distance does this attraction stop entirely?

The nearest body in space to the earth is the moon, some 240,000 miles away. Would an apple reach the earth if thrown from the moon? But perhaps the moon itself has attractive power? If so, since the apple would be much nearer the moon than the earth, the probabilities are that the apple would never reach the earth.

But hold! The apple is not the only object that falls to the ground. What is true of the apple is true of all other bodies—of all matter, large and small. Now there is the moon itself, a very large body. Does the earth exert any gravitational pull on the moon? To be sure, the moon is many thousands of miles away, but the moon is a very large body, and perhaps this size is in some way related to the power of attraction?

But then if the earth attracts the moon, why does not the moon fall to the earth?

A glance at the accompanying figure will help to answer this question. We must remember that the moon is not stationary, but travelling at tremendous speed—so much so, that it circles the entire earth every month. Now if the earth were absent the path of the moon would be a straight line, say *MB*. If, however, the earth exerts attraction, the moon would be pulled inward. Instead of following the line *MB* it would follow the curved path *MB'*. And again, the moon having arrived at B', is prevented from following the line $B'C$, but rather $B'C'$. So that the path instead of being a straight line tends to become curved. From Kepler's researches the probabilities were that this curve would assume the shape of an ellipse rather than a circle.

The only reason, then, why the moon does not fall to the earth is on account of its motion. Were it to stop moving even for the fraction of a second it would come straight down to us, and probably few would live to tell the tale.

Newton reasoned that what keeps the moon revolving around the earth is the gravitational pull of the latter. The next important step was to discover the law regulating this motion. Here Kepler's observations of the movements of the planets around the sun was of inestimable value; for from these Newton deduced the hypothesis that attraction varies inversely as the square of the distance. Making use of this hypothesis, Newton calculated what the attractive power possessed by the earth must be in order that the moon may continue in its path. He next compared this force with the force exerted by the earth in pulling the apple to the ground, and found the forces to be identical! "I compared," he writes, "the force necessary to keep the moon in her orb with the force of gravity at the surface of the earth, and found them answer pretty nearly." One and the same force pulls the moon and pulls the apple—the force of gravity. Further, the hypothesis that the force of gravity varies inversely as the square of the distance had now received experimental confirmation.

The next step was perfectly clear. If the moon's motion is controlled by the earth's gravitational pull, why is it not possible that the earth's motion, in turn, is controlled by the sun's gravitational pull? that, in fact, not only the earth's motion, but the motion of all the planets is regulated by the same means?

Here again Kepler's pioneer work was a foundation comparable to reinforced concrete. Kepler, as we have seen, had shown that the earth revolves around the sun in the form of an ellipse, one of the foci of this ellipse being occupied by the sun. Newton now proved that such an elliptic path was possible only if the intensity of the attractive force between sun and planet varied inversely as the square of the distance—the very same relationship that had been applied with such success in explaining the motion of the moon around the earth!

Newton showed that the moon, the sun, the planets—every body in space conformed to this law. The earth attracts the moon; but so does the moon the earth. If the moon revolves around the earth rather than the earth around the moon, it is because the earth is a much larger body, and hence its gravitational pull is stronger. The same is true of the relationship existing between the earth and the sun.

Further Developments of Newton's Law of Gravitation. When we speak of the earth attracting the moon, and the moon the earth, what we really mean is that every one of the myriad particles composing the earth attracts every one of the myriad particles composing the moon, and vice versa. If in dealing with the attractive forces existing between a planet and its satellite, or a planet and the sun, the power exerted by every one of these myriad particles would have to be considered separately, then the mathematical task of computing such forces might well appear hopeless. Newton was able to present the problem in a very simple form by pointing out that in a sphere such as the earth or the moon, the entire mass might be considered as residing in the center of the sphere. For purposes of computation, the earth can be considered a particle, with its entire mass concentrated at the center of the particle. This viewpoint enabled Newton to extend his law of inverse squares to the remotest bodies in the universe.

If this great law of Newton's found such general application beyond our planet, it served an equally useful purpose in explaining a number of puzzling features on this planet. The ebb and flow of the tides was one of these puzzles. Even in ancient times it had been noticed that a full moon and a high tide went hand in hand, and various mysterious powers, were attributed to the satellite and the ocean. Newton pointed out that the height of the water was a direct consequence of the attractive power of the moon, and, to a lesser extent, because further away, of the sun.

One of his first explanations, however, dealt with certain irregularities in the moon's motion around the earth. If the solar system would consist of the earth and moon alone, then the path of the moon would be that of an ellipse, with one of the foci of this ellipse occupied by the earth. Unfortunately for the simplicity of the problem, there are other bodies relatively near in space, particularly that huge body, the sun. The sun not only exerts its pull on the earth but also on the moon. However, as the sun is much further away from the moon than is the earth, the earth's attraction for its satellite is much greater, despite the fact that the sun is much huger and weighs far more than the earth. The greater pull of the earth in one direction, and a lesser pull of the sun in another, places the poor moon between the devil and the deep sea. The situation gives rise to a complexity of forces, the net result of which is that the moon's orbit is not quite that of an ellipse. Newton was able to account for all the forces that come into play, and he proved that the actual path of the moon was a direct consequence of the law of inverse squares in actual operation.

The "Principia." The law of gravitation, embodying also laws of motion, which we shall discuss presently, was first published in Newton's immortal "Principia" (1686). A selection from the preface will disclose the contents of the book, and, incidentally, the style of the author: "... We offer this work as mathematical principles of philosophy; for all the difficulty in philosophy seems to consist in this—from the phenomena of motions to investigate the forces of nature, and then from these forces to demonstrate the other phenomena; and to this end the general propositions in the first and second book are directed. In the third book we give an example of this in the explication of the system of the world; for by the propositions mathematically demonstrated in the first book, we there derive from the celestial phenomena the forces of gravity with which bodies tend to the sun and the several planets. Then, from these forces, by other propositions which are also mathematical, we deduce the motions of the planets, the comets, the moon and the sea. I wish we could derive the rest of the phenomena of nature by the same kind of reasoning from mechanical principles; for I am induced by many reasons to suspect that they may all depend upon certain forces by which the particles of bodies, by some causes hitherto unknown, are either mutually impelled towards each other, and cohere in regular figures, or are repelled and recede from each other...."

At this point we may state that neither Newton, nor any of Newton's successors including Einstein, have been able to advance even a plausible theory as to the nature of this gravitational force. We know that this force pulls a stone to the ground; we know, thanks to Newton, the laws regulating the motions due to gravity; but what this force we call gravity really is we do not know. The mystery is as deep as the mystery of the origin of life.

"Prof. Einstein," writes Prof. Eddington, "has sought, and has not reached, any ultimate explanation of its [that is, gravitation] cause. A certain connection between the gravitational field and the measurement of space has been postulated, but this throws light rather on the nature of our measurements than on gravitation itself. The relativity theory is indifferent to hypotheses as to the nature of gravitation, just as it is indifferent to hypotheses as to the nature of light."

Newton's Laws of Motion. In his *Principia* Newton begins with a series of simple definitions dealing with matter and force, and these are followed by his three famous laws of motion. The nature and amount of the effort required to start a body moving, and the conditions required to keep a body in motion, are included in these laws. The Fundamentals, mass, time and space, are exhibited in their various relationships. Of importance to us particularly is that in these laws, time and space are considered as definite entities, and as two distinct and widely separated manifestations. We shall see that in Einstein's hands a very close relationship between these two is brought about.

Both Newton and Einstein were led to their theory of gravitation by profound studies of the mathematics of motion, but as Newton's conception of motion differed from Einstein's, and as, moreover, important discoveries into the nature of matter and the relationship of motion to matter were made subsequent to Newton's time, we need not wonder that the two theories show divergence; that, as we shall see, Newton's is probably but an approximation of the truth. If we confine our attention to our own solar system, the deviation from Newton's law is, as a rule, so small as to be negligible.

Newton's laws of motion are really axioms, like the axioms of Euclid: they do not admit of direct proof; but there is this difference, that the axioms of Euclid seem more obviously true. For example, when Euclid informs us that "things which are equal to the same thing are equal to one another," we have no hesitation in accepting this statement, for it seems so self-evident. When, however, we are told by Newton that "the alteration of motion is ever proportional to the motive force impressed," we are at first somewhat bewildered with the phraseology, and then, even when that has been mastered, the readiness with which we respond will probably depend upon the amount of scientific training we have received.

"Every body continues in its state of rest or of uniform motion in a straight line, unless it is compelled to change that state by forces impressed thereon." So runs Newton's first law of motion. A body does not move unless something causes it to move; to make the body move you must overcome the *inertia* of the body. On the other hand, if a body is moving, it tends to continue moving, as witness our forward movement when the train is brought to a standstill. It may be asked, why does not a bullet continue moving indefinitely once it has left the barrel of the gun? Because of the resistance of the air which it has to overcome; and the path of the bullet is not straight because gravity acts on it and tends to pull it downwards.

We have no definite means of proving that a body once set in motion would continue moving, for an indefinite time, and along a straight line. What Newton meant was that a body would continue moving provided no external force acted on it; but in actual practise such a condition is unknown.

Newton's first law defines force as that action necessary to change a state of rest or of uniform motion, and tells us that force alone changes the motion of a body. His second law deals with the relation of the force applied and the resulting change of motion of the body; that is, it shows us how force may be measured. "The alteration of motion is ever proportional to the motive force impressed, and is made in the direction of the right line in which that force is impressed."

Newton's third law runs—"To every action there is always opposed an equal reaction." The very fact that you have to use force means that you have to overcome something of an opposite nature. The forward pull of a horse towing a boat equals the backward pull of the tow-rope connecting boat and horse. "Many people," says Prof. Watson, "find a difficulty in accepting this statement ... since they think that if the force exerted by the horse on the rope were not a little greater than the backward force exerted by the rope on the horse, the boat would not progress. In this case we must, however, remember that, as far as their relative positions are concerned, the horse and the boat are at rest, and form a single body, and the action and reaction between them, due to the tension on the rope, must be equal and opposite, for otherwise there would be relative motion, one with respect to the other."

It may well be asked, what bearing have these laws of Newton on the question of time and space? Simply this, that to measure force the factors necessary are the masses of the bodies concerned, the time involved and the space covered; and Newton's equations for measuring forces assume time and space to be quite independent of one another. As we shall see, this is in striking contrast to Einstein's view.

Newton's Researches on Light. In 1665, when but 23 years old, Newton invented the binomial theorem and the infinitesimal calculus, two phases of pure mathematics which have been the cause of many a sleepless night to college freshmen. Had Newton done nothing else his fame would have been secure. But we have already glanced at his law of inverse squares and the law of gravitation. We now have to turn to some of Newton's contributions to optics, because here more than elsewhere we shall find the starting point to a series of researches which have culminated so brilliantly in the work of Einstein.

Newton turned his attention to optics in 1666 when he proved that the light from the sun, which appears white to us, is in reality a mixture of all the colors of the rainbow. This he showed by placing a prism between the ray of light and a screen. A spectrum showing all the colors from red to violet appeared on the screen.

Another notable achievement of his was the design of a telescope which brought objects to a sharp focus and prevented the blurring effects which had occasioned so much annoyance to Newton and his predecessors in all their astronomical observations.

These and other discoveries of very great interest were brought together in a volume on optics which Newton published in 1704. Our particular concern here is to examine the views advanced by him as to the nature of light.

That the nature of light should have been a subject for speculation even to the ancients need not surprise us. If other senses, as touch, for example, convey impressions of objects, it is true to say that the sense of sight conveys the most complete impression. Our conception of the external world is largely based upon the sense of sight; particularly so when we deal with objects beyond our reach. In astronomy, therefore, a study of the properties of light is indispensable.[1]

But what is this light? We open our eyes and we see; we close our eyes and we fail to see. At night in a dark room we may have our eyes open and yet we do not see; light, then, must be absent. Evidently, light does not wholly depend upon whether our eyes are open or closed. This much is certain: the eye functions and something else functions. What is this "something else"?

Strangely enough, Plato and Aristotle regarded light as a property of the eye and the eye alone. Out of the eye tentacles were shot which intercepted the object and so illuminated it. From what has already been said, such a view seems highly unlikely. Far more consistent with their philosophy in other directions would have been the theory that light has its source in the object and not in the eye, and travels from object to eye rather than the reverse. How little substance the Aristotelian contribution possesses is immediately seen when we refer to the art of photography. Here light rays produce effects which are independent of any property of the eye. The blind man may click the camera and produce the impression on the plate.

Newton, of course, could have fallen into no such error as did Plato and Aristotle. The source of light to him was the luminous body. Such a body had the power of emitting minute particles at great speed, and these when coming in contact with the retina produce the sensation of sight.

This emission or corpuscular theory of Newton's was combated very strongly by his illustrious Dutch contemporary, Huyghens, who maintained that light was a wave phenomenon, the disturbance starting at the luminous body and spreading out in all directions. The wave motions of the sea offer a certain analogy.

Newton's strongest objection to Huyghens' wave theory was that it seemed to offer no satisfactory explanation as to why light travelled in straight lines. He says: "To me the fundamental supposition itself seems impossible, namely that the waves or vibrations of any fluid can, like the rays of light, be propagated in straight lines, without a continual and very extravagant bending and spreading every way into the quiescent medium, where they are terminated by it. I mistake if there be not both experiment and demonstration to the contrary."

In the corpuscular theory the particles emitted by the luminous body were supposed to travel in straight lines. In empty space the particles travelled in straight lines and spread in all directions. To explain how light could traverse some types of matter—liquids, for example—Newton supposed that these light particles travelled in the spaces between the molecules of the liquid.

Newton's objection to the wave theory was not answered very convincingly by Huyghens. Today we know that light waves of high frequency tend to travel in straight lines, but may be prevented from doing so by gravitational forces of bodies near its path. But this is Einstein's discovery.

A very famous experiment by Foucault in 1853 proved beyond the shadow of a doubt that Newton's corpuscular theory was untenable. According to Newton's theory, the velocity of light must be greater in a denser medium (such as water) than in a lighter one (such as air). According to the wave theory the reverse is true. Foucault showed that light does travel more slowly in water than in air. The facts were against Newton and in favor of Huyghens; and where facts and theory clash there is but one thing to do: discard the theory.

Some Facts about Newton. Newton was a Cambridge man, and Newton made Cambridge famous as a mathematical center. Since Newton's day Cambridge has boasted of a Clerk Maxwell and a Rayleigh, and her Larmor, her J. J. Thomson and her Rutherford are still with us. Newton entered Trinity College when he was 18 and soon threw himself into higher mathematics. In 1669, when but 27 years old, he became professor of mathematics at Cambridge, and later represented that seat of learning in Parliament. When his friend Montague became Chancellor of the Exchequer, Newton was offered, and accepted, the lucrative position of Master of the Mint. As president of the Royal Society Newton was occasionally brought in contact with Queen Anne. She held Newton in high esteem, and in 1705 she conferred the honor of knighthood on him. He died in 1727.

"I do not know," wrote Newton, "what I may appear to the world, but, to myself, I seem to have been only like a boy playing on the seashore, and directing myself in now and then finding a smoother pebble or a prettier shell than ordinary, whilst the great ocean of truth lay all undiscovered before me."

Such was the modesty of one whom many regard as the greatest intellect of all ages.

REFERENCES

An excellent account of Newton may be found in Sir R. S. Ball's *Great Astronomers* (Sir Isaac Pitman and Sons, Ltd., London). Sedgewick and Tyler's *A Short History of Science* (Macmillan, 1918) and Cajori's *A History of Mathematics* (Macmillan, 1917) may also be consulted to advantage. The standard biography is that by Brewster.

---

[1]

See Note 1 at the end of the volume. ↑

II

THE ETHER AND ITS CONSEQUENCES

Huyghens' wave theory of light, now so generally accepted, loses its entire significance if a medium for the propagation of these waves is left out of consideration. This medium we call the ether.[1]

Huyghens' reasoning may be illustrated in some such way as this: If a body moves a force pushes or pulls it. That force itself is exemplified in some kind of matter—say a horse. The horse in pulling a cart is attached to the cart. The horse in pulling a boat may not be attached to the boat directly but to a rope, which in turn is attached to the boat. In common cases where one piece of matter affects another, there is some direct contact, some go-between.

But cases are known where matter affects matter without affording us any evidence of contact. Take the case of a magnet's attraction for a piece of iron. Where is the rope that pulls the iron towards the magnet? Perhaps you think the attraction due to the air in between the magnet and iron? But removing the air does not stop the attraction. Yet how can we conceive of the iron being drawn to the magnet unless there is some go-between? some medium not readily perceptible to the senses perhaps, and therefore not strictly a form of matter?

If we can but picture some such medium we can imagine our magnet giving rise to vibrations in this medium which are carried to the iron. The magnet may give rise to a disturbance in that portion of the medium nearest to it; then this portion hands over the disturbance to its neighbor, the next portion of the medium; and so on, until the disturbance reaches the iron. You see, we are satisfying our sense-perception by arguing in favor of action by actual contact rather than some vague action at a distance; the go-between instead of being a rope is the medium called the ether.

Foucault's experiment completely shattered the corpuscular theory of light, and for want of any other more plausible alternative, we are thrown back on Huyghens' wave theory. It will presently appear that this wave theory has elements in it which make it an excellent alternative. In the meantime, if light is to be considered as a wave motion, then the query immediately arises, what is the medium through which these waves are propagated? If water is the medium for the waves of the sea, what is the medium for the waves of light? Again we answer, the medium is the ether.

What Is This "Ether"? Balloonists find conditions more and more uncomfortable the higher they ascend, for the density of the air (and therefore the amount of oxygen in a given volume of air) becomes less and less. Meteorologists have calculated that traces of the air we breathe may reach a height of some 200 miles. But what is beyond? Nothing but the ether, it is claimed. Light from the sun and stars reaches us via the ether.

But what is this ether? We cannot handle it. We cannot see it. It fails to fall within the scope of any of our senses, for every attempt to show its presence has failed. It is spirit-like in the popular sense. It is Lodge's medium for the souls of the departed.

Helmholtz and Kelvin tried to arrive at some properties of this hypothetical substance from a careful study of the manner in which waves were propagated through this ether. If, as the wave theory teaches us, the ether can be set in motion, then according to laws of mechanics, the ether has mass. If so it is smaller in amount than anything which can be detected with our most accurate balance. Further—and this is a difficulty not easily explained—if this ether has any mass, why does it offer no detectable resistance to the velocity of the planets in it? Why is not the velocity of the planets reduced in time, just as the velocity of a rifle bullet decreases owing to the resistance of the air?

Lodge, in arguing in favor of an ether, holds that its presence cannot be detected because it pervades all space and all matter. His favorite analogy is to point out the extreme unlikelihood of a deep-sea fish discovering the presence of the water with which it is surrounded on all sides;—all of which tells us nothing about the ether, but does try to tell us why we cannot detect it.[2]

In short, answering the query at the head of this paragraph, we may say that we do not know.

Waves Set up in This Ether. The waves are not all of the same length. Those that produce the sensation of sight are not the smallest waves known, yet their length is so small that it would take anywhere from one to two million of them to cover a yard. Curiously enough, our eye is not sensitive to wave lengths beyond either side of these limits; yet much smaller, and much larger waves are known. The smallest are the famous X-rays, which are scarcely one ten-thousandth the size of light waves. Waves which have a powerful chemical action—those which act on a photographic plate, for example—are longer than X-rays, yet smaller than light waves. Waves larger than light waves are those which produce the sensation of heat, and those used in wireless telegraphy. The latter may reach the enormous length of 5,000 yards. X-ray, actinic, or chemically active ray, light ray, heat ray, wireless ray—they differ in size, yet they all have this in common: they travel with the same speed (186,000 miles per second).

The Electromagnetic Theory of Light. Powerful support to the conception that space is pervaded by ether was given when Maxwell discovered light to be an electromagnetic phenomenon. From purely theoretical considerations this gifted English physicist was led to the view that waves could be set up as a result of electrical disturbances. He proved that such waves would travel with the same velocity as light waves. As air is not needed to transmit electrical phenomena—for you can pump all air out of a system and produce a vacuum, and electrical phenomena will continue—Maxwell was forced to the conclusion that the waves set up by electrical disturbances and transmitted with the same velocity as light, were enabled to do so with the help of the same medium as light, namely, the ether.

It was now but a step for Maxwell to formulate the theory that light itself is nothing but an electrical phenomenon, the sensation of light being due to the passage of electric waves through the ether. This theory met with considerable opposition at first. Physicists had been brought up in a school which had taught that light and electricity were two entirely unrelated phenomena, and it was difficult for them to loosen the shackles that bound them to the older school. But two startling discoveries helped to fasten attention upon Maxwell's theory. One was an experimental confirmation of Maxwell's theoretical deduction. Hertz, a pupil of Helmholtz, showed how the discharge from a Leyden jar set up oscillations, which in turn gave rise to waves in the ether, comparable, in so far as velocity is concerned, to light waves, but differing from the latter in wave length, the Hertzian waves being much longer. At a later date these waves were further investigated by Marconi, with the result that wireless messages soon began to be flashed from one place to another.

Just as there is a close connection between light and electricity, so there is between light and magnetism. The first to point out such a relationship was the illustrious Michael Faraday, but we owe to Zeeman the most extensive investigations in this field.

If we throw some common salt into a flame, and, with the help of a spectroscope, examine the spectrum produced, we are struck by two bright lines which stand out very prominently. These lines, yellow in color, are known as the D-lines and serve to identify even minute traces of sodium. What is true of sodium is true of other elements: they all produce very characteristic spectra. Now Zeeman found that if the flame is placed between a powerful magnet, and then some common salt thrown into the flame, the two yellow lines give place to ten yellow lines. Such is one of the results of the effect of a magnetic field on light.

The Electron. The "Zeeman effect" led to several theories regarding its nature. The most successful of these was one proposed by Larmor and more fully treated by Lorentz. It has already been pointed out that the only difference between wireless and light waves is that the former are much "longer," and, we may now add, their vibrations are much slower. Light and wireless waves bear a relationship to one another comparable to the relationship born by high and low-pitched sounds. To produce wireless waves we allow a charge of electricity to oscillate to and fro. These oscillations, or oscillating charges, are the cause of such waves. What charges give rise to light waves? Lorentz, from a study of the Zeeman effect, ascribed them to minute particles of matter, smaller than the chemical atom, to which the name "electron" was given.

The unit of electricity is the electron. Electrons in motion give rise to electricity, and electrons in vibration, to light. The Zeeman effect gave Lorentz enough data to calculate the mass of such electrons. He then showed that these electrons in a magnetic field would be disturbed by precisely the amount to which Zeeman's observations pointed. In other words, the assumption of the electron fitted in most admirably with Zeeman's experiments on magnetism and light.

In the meantime, a study of the discharge of electricity through gases, and, later, the discovery of radium, led, among other things, to a study of beta or cathode rays—negatively charged particles of electricity. Through a series of strikingly original experiments J. J. Thomson ascertained the mass of such particles or corpuscles, and then the very striking fact was brought out that Thomson's corpuscle weighed the same as Lorentz's electron. The electron was not merely the unit of electricity but the smallest particle of matter.

The Nature of Matter. All matter is made up of some eighty-odd elements. Oxygen, copper, lead are examples of such elements. Each element in turn consists of an innumerable number of atoms, of a size so small, that 300 million of them could be placed alongside of one another without their total length exceeding one inch.

John Dalton more than a hundred years ago postulated a theory, now known as the atomic theory, to explain one of the fundamental laws in chemistry. This theory started out with an old Greek assumption that matter cannot be divided indefinitely, but that, by continued subdivision, a point would be reached beyond which no further breaking up would be possible. The particles at this stage Dalton called atoms.

Dalton's atomic hypothesis became one of the pillars upon which the whole superstructure of chemistry rested, and this because it explained a number of perplexing difficulties so much more satisfactorily than any other hypothesis.

For nearly a century Dalton stood as firm as a rock. But early in the nineties some epoch-making experiments on the discharge of electricity through gases were begun by a group of physicists, particularly Crookes, Rutherford, Lenard, Roentgen, Becquerel, and, above all, J. J. Thomson, which pointed very clearly to the fact that the atoms are not the smallest particles of matter at all; that, in fact, they could be broken up into electrons, of a diameter one one-hundred-thousandth that of an atom.

It remained for the illustrious Madame Curie to confirm this beyond all doubt by her isolation of radium. Here, as

Madame Curie showed, was an element whose atoms were actually breaking up under one's very eyes, so to speak.

So far have we advanced since Dalton's day, that Dalton's unit, the atom, is now pictured as a complex particle patterned after our solar system, with a nucleus of positive electricity in the center, and particles of negative electricity, or electrons, surrounding the nucleus.

All this leads to one inevitable conclusion: matter is electrical in nature. But now if matter and light have the same origin, and matter is subject to gravitation, why not light also? So reasoned Einstein.

Summary. Newton's studies of matter in motion led to his theory of gravitation, and, incidentally, to his conception of time and space as definite entities. As we shall see, Einstein in his theory of gravitation based it upon discoveries belonging to the post-Newtonian period. One of these is Minkowski's theory of time and space as one and inseparable. This theory we shall discuss at some length in the next chapter.

Other important discoveries which led up to Einstein's work are the researches which culminated in the electron theory of matter. The origin of this theory may be traced to studies dealing with the nature of light.

Here again Newton appears as a pioneer. Newton's corpuscular theory, however, proved wholly untenable when Foucault showed that the velocity of light in water is less than in air, which is the very reverse of what the corpuscular theory demands, but which does agree with Huyghens' wave theory.

But Huyghens' wave theory postulated some medium in which the waves can act. To this medium the name "ether" was given. However, all attempts to show the presence of such an ether failed. Naturally enough, some began to doubt the existence of an ether altogether.

Huyghens' wave theory received a new lease of life with Maxwell's discovery that light is an electromagnetic phenomenon; that the waves set up by a source of light were comparable to waves set up by an electrical disturbance.

Zeeman next showed that magnetism was also, closely related to light.

A study of Zeeman's experiments led Lorentz to the conclusion that electrical phenomena are due to the motion of charged particles called "electrons," and that the vibrations of these electrons give rise to light.

The conception of the electron as the very fundamental of matter was arrived at in an entirely different way: from studies dealing with the discharge of electricity through gases and the breaking up of the atoms of radium.

If matter and light have the same origin, and if matter is subject to gravitation, why not light also?

REFERENCES

For the general subject of light the reader must be referred to a rather technical work, but one of the best in the English language: Edwin Edser, *Light for Students* (Macmillan, 1907).

The nature of matter and electricity is excellently discussed in several books of a popular variety. The very best and most complete of its kind that has come to the author's attention is Comstock and Troland's *The Nature of Matter and Electricity* (D. Van Nostrand Co., 1919). Two other very readable books are Soddy's *Matter and Energy* (Henry Holt and Co.) and Crehore's *The Mystery of Matter and Energy* (D. Van Nostrand Co., 1917).

1

See Note 2. ↑

2

See Note 3. ↑

ALBERT EINSTEIN

C. Wide World

III

EINSTEIN

"This is the most important result obtained in connection with the theory of gravitation since Newton's day. Einstein's reasoning is the result of one of the highest achievements of human thought."

These words were uttered by Sir J. J. Thomson, the president of the Royal Society, at a meeting of that body held on November 6, 1919, to discuss the results of the Eclipse Expedition.

Einstein another Newton—and this from the lips of J. J. Thomson, England's most illustrious physicist! If ever man weighed words carefully it is this Cambridge professor, whose own researches have assured him immortality for all time.

What has this Albert Einstein done to merit such extraordinary praise? With the world in turmoil, with classes and races in a death struggle, with millions suffering and starving, why do we find time to turn our attention to this

Jew? His ideas have no bearing on Europe's calamity. They will not add one bushel of wheat to starving populations.

The answer is not hard to find. Men come and men go, but the mystery of the universe remains. It is Einstein's glory to have given us a deeper insight into the universe. Our scientists are Huxley's agnostics: they do not deny activities beyond our planet; they merely center their attention on the knowable on this earth. Our philosophers, on the other hand, go far afield. Some of them soar so high that, like one poet's opinion of Shelley, the bubble bursts. Einstein, using the tools of the scientist—the experimentalist—built a skyscraper which ultimately reached the philosophical school. His rôle is the rôle of alcohol in causing water and ether (the anæsthetic) to mix. Ether and water will mix no better than oil and water, without the presence of alcohol; in its presence a uniform mixture is obtained.

The Object of the Eclipse Expedition. Einstein prophesied that a ray of light passing near the sun would be pulled out of its course, due to the action of gravity. He went even further. He predicted how much out of its course the ray would be deflected. This prediction was based on a theory of gravitation which Einstein had developed in great mathematical detail. The object of the British Eclipse Expedition was either to prove or disprove Einstein's assumption.

The Result of the Expedition. Einstein's prophecy was fulfilled almost to the letter.

The Significance of the Result. Since Einstein's theory of gravitation is intimately associated with certain revolutionary ideas concerning time and space, and, therefore, with

Fundamentals of the Universe, the net result of the expedition was to strengthen our belief in the validity of his new view of the universe.

It is our intention in the following pages to discuss the expedition and the larger aspects of Einstein's theory that follow from it. But before we do so we must have a clear idea of our solar system.

Our Solar System. In the center of our system is the sun, a flaming mass of fire, much bigger than our own earth, and very, very far away. The sun has its family of eight planets—of which the earth is one—which travel around the sun; and around some of the planets there travel satellites, or moons. The earth has such a satellite, *the* moon.

Now we have good reasons for believing that every star which twinkles in the sky is a sun comparable to our own, having also its own planets and its own moons. These stars, or suns, are so much further away from us than our own sun, that but a speck of their light reaches us, and then only at night, when, as the poets would say, our sun has gone to its resting place.

The distances between bodies in the solar system is so immense that, like the number of dollars spent in the Great War, the number of miles conveys little, or no impression. But picture yourself in an express train going at the average rate of 30 miles an hour. If you start from New York and travel continuously you would reach San Francisco in 4 days. If you could continue your journey around the earth at the same rate you would complete it in 35 days. If now you could travel *into* space and *to* the moon, still with the same velocity, you would reach it in 350 days. Having reached the moon, you could circumscribe it with the same express train in 8 days, as compared to the 35 days it would take you to circumscribe the earth. If instead of travelling to the moon you would book your passage for the sun you, or rather your descendants, would get there in 350 *years*, and it would then take them 10 additional years to travel around the sun.

Immense as these distances are, they are small as compared to the distances that separate us from the stars. It takes light which, instead of travelling 30 miles an hour, travels 186,000 miles a *second*, about 8 minutes to get to us from the sun, and a little over 4 years to reach us from the nearest star. The light from some of the other stars do not reach us for several hundred years.

The Eclipse of the Sun. Now to return to an infinitesimal part of the universe—our solar system. We have seen that the earth travels around the sun, and the moon around the earth. At some time in the course of these revolutions the moon must come directly between the earth and the sun. Then we get the *eclipse* of the sun. As the moon is smaller than the earth, only a portion of the earth's surface will be cut off from the sun's rays. That portion which is so cut off suffers a total eclipse. This explains why the eclipse of May, 1919, which was a total one for Brazil, was but a partial one for us.

Einstein's Assertion Restated. Einstein claimed that a ray of light from one of the stars, if passing near enough to the surface of the sun, would be appreciably deflected from its course; and he calculated the exact amount of this deflection. To begin with, why should Einstein suppose that the path of a ray of light would be affected by the son?

Newton's law of gravitation made it clear that bodies which have mass attract one another. If light has mass—and very recent work tends to show that it has—there is no reason why light should not be attracted by the sun, or any other planetary body. The question that agitated scientists was not so much whether a ray of light would be deviated from its path, but to what extent this deviation would take place. Would Einstein's figures be confirmed?

Of the bodies within our solar system the sun is by far the largest, and therefore it would exert a far greater pull than any of the planets on light rays coming from the stars. Under ordinary conditions, however, the sun itself shines with such brilliancy, that objects around it, including rays of light passing near its surface, are wholly dimmed. Hence the necessity of putting our theory to the test only when the moon covers up the sun—when there is a total eclipse of the sun.

A Graphical Representation. Imagine a star A, so selected that as its light comes to us the ray just grazes the sun. If the path of the ray is straight—if the sun has no influence on it—then the path can be represented by the line AB. If, however, the sun does exert a gravitational pull, then its real path will be AB', and to an observer on the earth the star will have appeared to shift from A to A'.

What the Eclipse Expedition Set Out to Do. Photographs of stars around the sun were to be taken during the eclipse, and these photographs compared with others of the same region taken at night, with the sun absent. Any apparent shifting of the stars could be determined by measuring the distances between the stars as shown on the photographic plates.

Three Possibilities Anticipated. According to Newton's assumption, light consists of corpuscles, or minute particles, emitted from the source of light. If this be true these particles, having mass, should be affected by the gravitational pull of the sun. If we apply Newton's theory of gravitation and make use of his formula, it can be shown that such a gravitational pull would displace the ray of light by an average amount equal to 0.75 (seconds of angular distance.)[1] On the other hand, where light is regarded as waves set in motion in the "ether" of space (the wave theory of light), and where weight is denied light altogether, no deviation need be expected. Finally there is a third alternative: Einstein's. Light, says Einstein, has mass, and therefore probably weight. Mass is the matter light contains; weight represents pull by gravity. Light rays will be attracted by the sun, but according to Einstein's theory of gravitation the sun's gravitational pull will displace the rays by an average amount equal to 1.75 (seconds of angular distance).

The Expeditions. That science is highly international, despite many recent examples to the contrary, is evidenced by this British Eclipse Expedition. Here was a theory propounded by one who had accepted a chair of physics in the university of Berlin, and across the English Channel were Germany's mortal enemies making elaborate preparations to test the validity of the Berlin professor's theory.

The British Astronomical Society began to plan the eclipse expedition even before the outbreak of the Great War. During the years that followed, despite the destinies of nations which hung on threads from day to day, despite the darkest hours in the history of the British people, our English astronomers continued to give attention to the details of the proposed expedition. When the day of the eclipse came all was in readiness.

One expedition under Dr. Crommelin was sent to Sobral, Brazil; another, under Prof. Eddington, to Principe, an island off the west coast of Africa. In both these places a total eclipse was anticipated.

The eclipse occurred on May 29, 1919. It lasted for six to eight minutes. Some 15 photographs, with an average exposure of five to six seconds, were taken. Two months later another series of photographs of the same region were taken, but this time the sun was no longer in the midst of these stars.

The photographs were brought to the famous Greenwich Observatory, near London, and the astronomers and mathematicians began their laborious measurements and calculations.

On November 6, at the meeting of the Royal Society, the result was announced. The Sobral expedition reported 1.98; the Principe expedition 1.62. The average was 1.8. Einstein had predicted 1.75, Newton might have predicted 0.75, and the orthodox scientists would have predicted 0. There could now no longer be any question as to which of the three theories rested on a sure foundation. To quote Sir Frank Dyson, the Astronomer Royal: "After a careful study of the plates I am prepared to say that there can be no doubt that they confirm Einstein's prediction. A very definite result has been obtained that light is deflected in accordance with Einstein's law of gravitation."[2]

Where Did Einstein Get His Idea of Gravitation? In 1905 Einstein published the first of a series of papers supporting and extending a theory of time and space to which the name "the theory of relativity" had been given. These views as expounded by Einstein came into direct conflict with Newton's ideas of time and space, and also with Newton's law of gravitation. Since Einstein had more faith in his theory of relativity than in Newton's theory of gravitation, Einstein so changed the latter as to make it harmonize with the former. More will be said on this subject.

Let not the reader misunderstand. Newton was not wholly in the wrong; he was only approximately right. With the knowledge existing in Newton's day Newton could have done no more than he did; no mortal could have done more. But since Newton's day physics—and science in general—has advanced in great strides, and Einstein can interpret present-day knowledge in the same masterful fashion that Newton could in his day. With more facts to build upon, Einstein's law of gravitation is more universal than Newton's; it really includes Newton's.

But now we must turn our attention very briefly to the theory of relativity—the theory that led up to Einstein's law of gravitation.

The Theory of Relativity. The story goes that Einstein was led to his ideas by watching a man fall from a roof. This story bears a striking similarity to Newton and the apple. Perhaps one is as true as the other.[3]

However that may be, the principle of relativity is as old as philosophical thought, for it denies the possibility of measuring absolute time, or absolute space. All things are relative. We say that it takes a "long time" to get from New York to Albany; long as compared to what? long, perhaps, as compared to the time it takes to go from New York City to

37

Brooklyn. We say the White House is "large"; large when compared to a room in an apartment. But we can just as well reverse our ideas of time and distance. The time it takes to go from New York to Albany is "short" when compared to the time it takes to go from New York to San Francisco. The size of the White House is "small" when compared to the size of the city of Washington.

Let us take another illustration. Every time the earth turns on its axis we mark down as a day. With this as a basis, we say that it takes a little over 365 days for the earth to complete its revolution around the sun, and our 365 days we call a year. But now consider some of our other planets. With our time as a basis, it takes Jupiter or Saturn 10 hours to turn on its axis, as compared to the 24 hours it takes the earth to turn. Saturn's day is less than one-half our day, and our day is more than twice Saturn's—that is, according to the calculations of the inhabitants of the earth. Mercury completes her circuit around the sun in 88 days; Neptune, in 164 years. Mercury's year is but one-fourth ours, Neptune's, 164 times ours. And observers at Mercury and Neptune would regard us from *their* standard of time, which differs from our standard.

You may say, why not take *our* standard of time as *the* standard, and measure everything by it? But why should you? Such a selection would be quite arbitrary. It would not be based on anything absolute, but would merely depend on our velocity around the sun.

These ideas are old enough in metaphysics. Einstein's improvement of them consists not merely in speculating about them, but in giving them mathematical form.

The Origin of the Theory of Relativity. A train moves with reference to the earth. The earth moves with reference to the sun. We say the sun is stationary and the earth moves around the sun. But how do we know that the sun itself does not move with reference to some other body? How do we know that our planetary system, and the stars, and the cosmos as a whole is not in motion?

There is no way of answering such a question unless we could get a point of reference which is fixed—fixed absolutely in space.

We have already alluded to our view of the nature of light, known as the wave theory of light. This theory postulates the existence of an all-pervading "ether" in space. Light sets up wave disturbances in this ether, and is thus propagated. If the ocean were the ether, the waves of the ocean would compare with the waves set up by the ether.

But what is this ether? It cannot be seen. It defies weight. It permeates all space. It permeates all matter. So say the exponents of this ether. To the layman this sounds very much like another name for the Deity. To Sir Oliver Lodge it represents the spirits of the departed.

To us, of importance is the conception that this ether is absolutely stationary. Such a conception is logical if the various developments in optics and electricity are considered. But if absolutely stationary, then the ether is the long-sought-for point of reference, the guide to determine the motion of all bodies in the universe.

The Famous Experiment Performed by Prof. Michelson. If there is an ether, and a stationary ether, and if the earth moves with reference to this ether, the earth, in moving, must set up ether "currents"—just as when a train moves it sets up air currents. So reasoned Michelson, a young Annapolis graduate at the time. And forthwith he devised a crucial experiment the explanation of which we can simplify by the following analogy:

Which is the quicker, to swim up stream a certain length, say a hundred yards, and back again, or across stream the same length and back again? The swimmer will answer that the up-and-down journey is longer.[4]

Our river is the ether. The earth, if moving in this ether, will set up an ether stream, the up stream being parallel to the earth's motion. Now suppose we send a beam of light a certain distance up this ether stream and back, and note the time; and then turn the beam of light at right angles and send it an equal distance across the stream and back, and note the time. How will the time taken for light to travel in these two directions compare? Reasoning by analogy, the up-and-down stream should take longer.

Michelson's results did not accord with analogy. No difference in time could be detected between the beam of light travelling up-and-down, and across-and-back.

But this was contrary to all reason if the postulate of an ether was sound. Must we then revise our ideas of an ether? Perhaps after all there is no ether.

But if no ether, how are we to explain the propagation of light in space, and various electrical phenomena connected with it, such as the Hertzian, or wireless waves?

There was another alternative, one suggested by Larmor in England and Lorentz in Holland,—that matter is contracted in the direction of its motion through the ether current. To say that bodies are actually shortened in the direction of their motion—by an amount which increases as the velocity of these bodies approaches that of light—is so revolutionary an idea that Larmor and Lorentz would hardly have adopted such a viewpoint but for the fact that recent investigations into the nature of matter gave basis for such belief.

Matter, it has been shown, is electrical in nature. The forces which hold the particles together are electrical. Lorentz showed that mathematical formulas for electrical forces could be developed which would inevitably lead to the view that material bodies contract in the direction of their motion.[5]

"But this is ridiculous," you say; "if I am shorter in one direction than in another I would notice it." You would if some things were shortened and others were not. But if all things pointing in a certain direction are shortened to an equal extent, how are you going to notice it? Will you apply the yard stick? That has been shortened. Will you pass judgment with the help of your eyes? But your retina has also contracted. In brief, if all things contract to the same amount it is as if there were no contraction at all.

Lorentz's Plausible Explanation Really Deepens the Mystery. The startling ideas just outlined have opened up several new vistas, but they have left unanswered the two problems we set out to solve: whether there is an ether, and if so, what is the velocity of the earth in reference to this ether? Lorentz maintains that there is an ether, but the velocities of bodies relative to it must forever remain a mystery. As you change your position your distances change; you change; everything about you changes accordingly; and all basis for comparison is lost. Nature has entered into a conspiracy to keep you ignorant.

Einstein Comes upon the Scene. Einstein starts with the assumption that there is no possible way of identifying this ether. Suppose we ignore the ether altogether, what then?[6]

If we do ignore the ether we no longer have any absolute point of reference; for if the ether is considered stationary the velocity of all bodies within the ether may be referred to it; any point in space may be considered a fixed point. If, however, there is no ether, or if we are to ignore it, how are we to get the velocity of bodies in space?

The Principle of Relativity. If we are to believe in the "causal relationship between only such things as lie within the realm of observation," then *observation* teaches us that bodies move only *relative* to one another, and that the idea of *absolute* motion of a body in space is meaningless. Einstein, therefore, postulates that there is no such thing as absolute motion, and that all we can discuss is the relative motion of one body with respect to another. This is just as logical a deduction from Michelson's experiment as the attempt to explain Michelson's anomalous results in the light of an all-pervading ether.

Consider for a moment Newton's scheme. This great pioneer pictured an absolute standard of position in space relative to which all velocities are measured. Velocities were measured by noting the distance covered and dividing the result by the time taken to cover the distance. Space was a definite entity; and so was time. "Time," said Newton, "flows evenly on," independent of aught else. To Newton time and space were entirely different, in no way to be confounded.

Just as Newton conceived of absolute space, so he conceived of absolute time. From the latter standard of reference the idea of a "simultaneity of events" at different places arose. But now if there is no standard of reference, if the ether does not exist or does not function, if two points A and B cannot be referred to a third, and fixed point C, how can we talk of "simultaneity of events" at A and B?

In fact, Einstein shows that if all you can speak about is relative motion, then one event which takes say one minute on one planet would not take one minute on another. For consider two bodies in space, say the planets Venus and the earth, with an observer B on Venus and another A on the earth. B notes the time taken for a ray of light to travel from B to the distance M. A on the earth has means of observing the same event. B records one minute. A is puzzled, for his watch records a little more than one minute. What is the explanation? Granting that the two clocks register the same time to start with, and assuming further Einstein's hypothesis that the velocity of light is independent of its source, the difference in time is due to the fact that the planet Venus moves with reference to the observer on the earth; so that A in reality does not measure the path BM and MB, but BM' and $M'B'$, where BB' represents the distance Venus itself has moved in the interval. And if you put yourself in B's position on Venus the situation is exactly reversed. All of which is simply another way of saying that what is a certain time on one body in space is another time on another body in space. There is nothing definite in time.

Prof. Cohen's Illustration. Further bewildering possibilities are clearly outlined in this apt illustration: "If when you are going away on a long and continuous journey you write home at regular intervals, you should not be surprised that with the best possible mail service your letters will reach home at progressively longer intervals, since each letter will have a greater distance to travel than its predecessor. If you were armed with instruments to hear the home clock ticking, you would find that as your distance from home keeps on increasing, the intervals between the successive ticks (that is, its seconds) grow longer, so that if you travelled with the velocity of sound the home clock would seem to slow down to a standstill—you would never hear the next tick.

"Precisely the same is true if you substitute light rays for sound waves. If with the naked eye or with a telescope you watch a clock moving away from you, you will find that its minute hand takes a longer time to cover its five-minute intervals than does the chronometer in your hand, and if the clock travelled with the velocity of light you would forever see the minute hand at precisely the same point. That which is true of the clock is, of course, also true of all time intervals which it measures, so that if you moved away from the earth with the velocity of light everything on it would appear as still as on a painted canvas."

Your time has apparently come to a standstill in one position and is moving in another! All this seems absurd enough, but it does show that time alone has little meaning.

Minkowski's Conclusion. The relativity theory requires that we thoroughly reorganise our method of measuring time. But this is intimately associated with our method of measuring space, the distance between two points. As we proceed we find that space without time has little meaning, and vice versa. This leads Minkowski to the conclusion that "time by itself and space by itself are mere shadows; they are only two aspects of a single and indivisible manner of coordinating the facts of the physical world." Einstein incorporated this time-space idea in his theory of relativity.

How We Measure a Point in Space. Suppose I say to you that the chemical laboratory of Columbia University faces Broadway; would that locate the laboratory? Hardly, for any building along Broadway would face Broadway. But suppose I add that it is situated at Broadway and 117th Street, southeast? there could be little doubt then. But if, further, this laboratory would occupy but part of the building, say the third floor; then the situation would be specified by naming Broadway, 117th Street S. E., third floor. If Broadway represents length, 117th Street width, and third floor height, we can see what is meant when we say that three dimensions are required to locate a position in space.

The Fourth Dimension. A point on a line may be located by one dimension; a point on a wall requires two dimensions; a point in the room, like the chemical laboratory above ground, needs three. The layman cannot grasp the meaning of a fourth dimension; yet the mathematician does imagine it, and plays with it in mathematical terms. Minkowski and Einstein picture time as the fourth dimension. To them time occupies no more important position than length, breadth, or thickness, and is as intimately related to these three as the three are to one another. H. G. Wells, the novelist, has beautifully caught this spirit when in his novel, "The Time Machine," he makes his hero travel backwards and forwards along time just as a man might go north or south. When the man with his time machine goes forward he is in the future; when he goes backwards he is in the past.

In reality, if we stop to think a minute, there is no valid reason for the non-existence of a fourth dimension. If one, two and three dimensions, why not four—and five and six, for that matter? Theoretically at least there is no reason why the limit should be set at three. However, our minds become sluggish when we attempt to picture dimensions beyond three; just as an extraordinary effort on our part is needed to follow Einstein when he "juggles" with space and time.

Our difficulty in imagining four dimensions may be likened to the difficulty two-dimensioned beings would experience in imagining us, beings of the conventional three dimensions. Suppose these two-dimensional beings were living on the surface of the earth; what could they see? They could see nothing below and nothing above the surface. They would see shifting surfaces as we walked about, but being sensitive to length and breadth only, and not to height, they could gain no notion whatsoever of what we really look like. It is thus with us when we attempt to picture four-dimensional space.

Perhaps the analogy of the motion picture may help us somewhat. As everybody knows, these motion pictures consist of a series of photographs which are shown in rapid succession on the screen. Each photograph by itself conveys a sensation of space, that is, of three dimensions; but one photograph rapidly following another conveys the sensation of space and time—four dimensions. Space and time are interlinked.

The Time-space Idea Further Developed. We have already alluded to the fact that objects in space moving with different velocities build up different time intervals. Thus the velocity of the star Arcturus, if compared with reference to the earth, moves at the rate of 200 miles a second. Its motion through space is different from ours. Objects which, according to Lorentz, contract in the direction of their motion to an extent proportional to their velocity, will contract differently on the surface of Arcturus than on the earth. Our space is not Arcturus' space; neither is Arcturus' time our time. And what is true of the discrepancies existing between the space and time conceptions of the earth and Arcturus is true of any other two bodies in space moving at different velocities.

But is there no relationship existing between the space and time of one body in the universe as compared to the space and time of another? Can we not find *something* which holds good for all bodies in the universe? We can. We can express it mathematically. It is the concept of time and space interlinked; of time as the fourth dimension, length, breadth and thickness being the other three; of time as one of four coordinates and at right angles to the other three (a situation which requires a terrific stretch of the imagination to visualize). The four dimensions are sufficient to co-ordinate the time-space relationships of all bodies in the cosmos, and hence have a universality which is totally lacking when time and space are used independently of one another. The four components of our time-space are up-and-down, right-and-left, backwards-and-forwards, and *sooner-and-later*.

"*Strain*" and "*Distortion*" in *Space*. The four-dimensional unit has been given the name "world-line," for the "world-line" of any particle in space is in reality a complete history of that particle as it moves about in space. Particles, we know, attract one another. If each particle is represented by a world-line these world-lines will be deflected from their course owing to such attraction.

Imagine a bladder representing the universe, with lines on it representing world-lines. Now squeeze the bladder. The world-lines are bent in various directions; they are "distorted." This illustrates the influence of gravity on these world-lines; it is the "strain" brought about due to the force of attraction. The distorted bladder illustrates even more, for it is a true representation of the real world.

How Einstein's Conception of Time and Space Led to a New View of Gravitation. In our conventional language we speak of the sun as exerting a "force" on the earth. We have seen, however, that this force brings about a "distortion" or "strain" in world-lines; or, what amounts to the same thing, a "distortion" or "strain" of time and space. The sun's "force," the "force" of any body in space, is the "force" due to gravity; and these "forces" may now be treated in terms of the laws of time and space. "The earth," Prof. Eddington tells us, "moves in a curved orbit, not because the sun exerts any direct pull, but because the earth is trying to find the shortest way through a space and time which have been tangled up by an influence radiating from the sun."[8]

At this point Newton's conceptions fail, for his views and his laws do not include "strains" in space. Newton's law of gravitation must be supplanted by one which does include such distortions. It is Einstein's great glory to have supplied us with this new law.

Einstein's Law of Gravitation. This appears to be the only law which meets all requirements. It includes Newton's law, and cannot be distinguished from it if our experiments are confined to the earth and deal with relatively small velocities. But when we betake ourselves to some orbits in space, with a gravitational pull much greater than the earth's, and when we deal with velocities comparable to that of light, the differences become marked.

Einstein's Theory Scores Its First Great Victory. In the beginning of this chapter we referred to the elaborate eclipse expedition sent by the British to test the validity of Einstein's new theory of gravitation. The British scientists would hardly have expended so much time and energy on this theory of Einstein's but for the fact that Einstein had already scored one great victory. What was it?

Imagine but a single planet revolving about the sun. According to Newton's law of gravitation, the planet's path would be that of an ellipse—that is, oval—and the planet would travel indefinitely along this path. According to Einstein the path would also be elliptical, but before a revolution would be quite completed, the planet would start along a slightly advanced line, forming a new ellipse slightly in advance of the first. The elliptic orbit slowly turns in the direction in which the planet is moving. After many years—centuries—the orbit will be in a different direction.

The rapidity of the orbit's change of direction depends on the velocity of the planet. Mercury moving at the rate of 30 miles a second is the fastest among the planets. It has the further advantage over Venus or the earth in that its orbit, as we have said, is an ellipse, whereas the orbits of Venus and the earth are nearly circular; and how are you going to tell in which direction a circle is pointing?

Observation tells us that the orbit of Mercury is advancing at the rate of 574 seconds (of arc) per century. We can calculate how much of this is due to the gravitational influence of other planets. It amounts to 532 seconds per century. What of the remaining 42 seconds?

You might be inclined to attribute this shortcoming to experimental error. But when all such possibilities are allowed for our mathematicians assure us that the discrepancy is 30 times greater than any possible experimental error.

This discrepancy between theory and observation remained one of the great puzzles in astronomy until Einstein cleared up the mystery. According to Einstein's theory the mathematics of the situation is simply this: in one revolution of the planet the orbit will advance by a fraction of a revolution equal to three times the square of the ratio of the velocity of the planet to the velocity of light. When we allow mathematicians to work this out we get the figure 43, which is certainly close enough to 42 to be called identical with it.

Still Another Victory? Einstein's third prediction—the shifting of spectral lines toward the red end of the spectrum in the case of light coming to us from the stars of appreciable mass—seems to have been confirmed recently (March, 1920). "The young physicists in Bonn," writes Prof. Einstein to a friend, "have now as good as certainty (*so gut wie sicher*) proved the red displacement of the spectral lines and have cleared up the grounds of a previous disappointment."

Summary. Velocity, or movement in space, is at the basis of Einstein's work, as it was at the basis of Newton's. But time and space no longer have the distinct meanings that they had when examined with the help of Newton's equations. Time and space are not independent but interdependent. They are meaningless when treated as separate entities, giving results which may hold for one body in the universe but do not hold for any other body. To get general laws which are applicable to the cosmos as a whole the Fundamentals of Mechanics must be united.

Einstein's great achievement consists in applying this revised conception of space and time to elucidate cosmical problems. "World-lines," representing the progress of particles in space, consisting of space-time combinations (the four dimensions), are "strained" or "distorted" in space due to the attraction that bodies exhibit for one another (the force of gravitation). On the other hand, gravitation itself—more universal than anything else in the universe—may be interpreted in terms of strains on world-lines, or, what amounts to the same thing, strains of space-time combinations. This brings gravitation within the field of Einstein's conception of time and space.

That Einstein's conception of the universe is an improvement upon that of Newton's is evidenced by the fact that Einstein's law explains all that Newton's law does, and also other facts which Newton's law is incapable of explaining. Among these may be mentioned the distortion of the oval orbits of planets round the sun (confirmed in the case of the planet Mercury), and the deviation of light rays in a gravitational field (confirmed by the English Solar Eclipse Expedition).

Einstein's Theories and the Inferences to be Drawn from Them. Einstein's theories, supported as they are by very convincing experiments, will probably profoundly influence philosophic and perhaps religious thought, but they can hardly be said to be of immediate consequence to the man in the street. As I have said elsewhere, Einstein's theories are not going to add one bushel of wheat to war-torn and devastated Europe, but in their conception of a cosmos decidedly at variance with anything yet conceived by any school of philosophy, they will attract the universal attention of thinking men in all countries. The scientist is immediately struck by the way Einstein has utilized the various achievements in physics and mathematics to build up a coordinated system showing connecting links where heretofore none were perceived. The philosopher is equally fascinated with a theory, which, in detail extremely complex, shows a singular beauty of unity in design when viewed as a whole. The revolutionary ideas propounded regarding time and space, the brilliant way in which the most universal property of matter, gravitation, is for the first time linked up with other properties of matter, and, above all, the experimental confirmation of several of his more startling predictions—always the finest test of scientific merit—stamps Einstein as one of those super-men who from time to time are sent to us to give us a peep into the beyond.

Some Facts about Einstein Himself. Albert Einstein was born in Germany some 45 years ago. At first he was engaged at the Patent Bureau in Berne, and later became professor at the Zürich Polytechnic. After a short stay at Prague University he accepted one of those tempting "Akademiker" professorships at the university of Berlin—professorships which insure a comfortable income to the recipient of one of them, little university work beyond, perhaps, one lecture a week, and splendid facilities for research. A similar inducement enticed the chemical philosopher, Van 't Hoff, to leave his Amsterdam, and the Swedes came perilously near losing their most illustrious scientist, Arrhenius.

Einstein published his first paper on relativity in 1905, when not more than 30 years old. Of this paper Planck, the Nobel Laureate in physics this year, has offered this opinion: "It surpasses in boldness everything previously suggested in speculative natural philosophy and even in the philosophical theories of knowledge. The revolution introduced into the physical conceptions of the world is only to be compared in extent and depth with that brought about by the introduction of the Copernican system of the universe."

Einstein published a full exposition of the relativity theory in 1916.

During the momentous years of 1914–19, Einstein quietly pursued his labors. There seems to be some foundation for the belief that the ways of the German High Command found little favor in his eyes. At any rate, he was not one of the forty professors who signed the famous manifesto extolling Germany's aims. "We know for a fact," writes Dr. O. A. Rankine, of the Imperial College of Science and Technology, London, "that Einstein never was employed on war work. Whatever may have been Germany's mistakes in other directions, she left her men of science severely alone. In fact, they were encouraged to continue in their normal occupations. Einstein undoubtedly received a large measure of support from the Imperial Government, even when the German armies were being driven back across Belgium."

Quite recently (June, 1920) the *Barnard* Medal of Columbia University was conferred on him "in recognition of his highly original and fruitful development of the fundamental concepts of physics through application of mathematics." In acknowledging the honor, Prof. Einstein wrote to President Butler that "... quite apart from the personal satisfaction, I believe I may regard your decision [to confer the medal upon him] as a harbinger of a better time in which a sense of international solidarity will once more unite scholars of the various countries."

References

For those lacking all astronomical knowledge, an excellent plan would be to read the first 40 pages of W. H. Snyder's *Everyday Science* (Allyn and Bacon), in which may be found a clear and simple account of the solar system. This could be followed with Bertrand Russell's chapter on *The Nature of Matter* in his little volume, *The Problems of Philosophy* (Henry Holt and Co.). Here the reader will be introduced to the purely philosophical side of the question—quite a necessary equipment for the understanding of Einstein's theory.

Of the non-mathematical articles which have appeared, those by Prof. A. S. Eddington (*Nature*, volume 101, pages 15 and 34, 1918) and Prof. M. R. Cohen (*The New Republic*, Jan. 21, 1920) are the best which have come to the author's notice. Other articles on Einstein's theory, some easily comprehensible, others somewhat confusing, and still others full of noise and rather empty, are by H. A. Lorentz, *The New York Times*, Dec. 21, 1919 (since reprinted in book form by Brentano's, New York, 1920); J. Q. Stewart, *Scientific American*, Jan. 3, 1920; E. Cunningham, *Nature*, volume 104, pages 354 and 374, 1919; F. H. Loring, *Chemical News*, volume 112, pages 226, 236, 248, and 260, 1915; E. B. Wilson, *Scientific Monthly*, volume 10, page 217, 1920; J. S. Ames, *Science*, volume 51, page 253, 1920^2; L. A. Bauer, *Science*, volume 51, page 301 (1920), and volume 51, page 581 (1920); Sir Oliver Lodge, *Scientific Monthly*, volume 10, page 378, 1920; E. E. Slosson, *Independent*, Nov. 29, Dec.

13, Dec. 20, Dec. 27, 1919 (since collected and published in book form by Harcourt, Brace and Howe); Isabel M. Lewis, *Electrical Experimenter*, Jan., 1920; A. J. Lotka, *Harper's Magazine*, March, 1920, page 477; and R. D. Carmichael, *New York Times*, March 28, 1920. Einstein himself is responsible for a brief article in English which first appeared in the London *Times*, and was later reprinted in *Science*, volume 51, page 8, 1920 (see the Appendix).

A number of books deal with the subject, and all of them are more or less mathematical. However, in every one of these volumes certain chapters, or portions of chapters, may be read with profit even by the non-mathematical reader. Some of these books are: Erwin Freundlich, *The Foundations of Einstein's Theory of Gravitation* (University Press, Cambridge, 1920). (A very complete list of references—up to Feb., 1920—is also given); A. S. Eddington, Report on the Relativity Theory of Gravitation for the Physical Society of London (Fleetway Press, Ltd., London, 1920); R. C. Tolman, *Theory of the Relativity of Motion* (University of California Press, 1917); E. Cunningham, *Relativity and the Electron Theory* (Longmans, Green and Co., 1915); R. D. Carmichael, *The Theory of Relativity* (John Wiley and Sons, 1913); L. Silberstein, *The Theory of Relativity* (Macmillan, 1914); and E. Cunningham, *The Principle of Relativity* (University Press, Cambridge, England, 1914).

To those familiar with the German language Einstein's book, *Über die spezielle und die allgemeine Relativitätstheorie* (Friedr. Vieweg und Sohn, Braunschweig, 1920), may be recommended.[10]

The mathematical student may be referred to a volume incorporating the more important papers of Einstein,

Minkowski and Lorentz: *Das Relativitätsprinzip*, (B. G. Teubner, Berlin, 1913).

Einstein's papers have appeared in the *Annalen der Physik*, Leipzig, volume 17, page 132, 1905, volume 49, page 760, 1916, and volume 55, page 241, 1918.

1

A circle—in our case the horizon—is measured by dividing the circumference into 360 parts; each part is called a degree. Each degree is divided into 60 minutes, and each minute into 60 seconds. ↑

2

See page 113. ↑

3

See Note 4. ↑

4

See Note 5. ↑

5

See Note 6. ↑

6

See Note 7. ↑

7

See Note 8. ↑

8

See Note 9. ↑

9

See page 93. ↑

10

This has since been translated into English by Dr. Lawson and published by Methuen (London).

Since the above has been written two excellent books have been published. One is by Prof. A. S. Eddington, *Space, Time and Gravitation* (Cambridge Univ. Press, 1920). The other, somewhat more of a philosophical work, is Prof. Moritz Schlick's *Space and Time in Contemporary Physics* (Oxford Univ. Press, 1920).

Though published as early as 1897, Bertrand Russell's *An Essay on the Foundations of Geometry* (Cambridge Univ. Press, 1897) contains a fine account of non-Euclidean geometry. ↑

APPENDIX

Note 1 (page 21)

"On this earth there is indeed a tiny corner of the universe accessible to other senses [than the sense of sight]: but feeling and taste act only at those minute distances which separate particles of matter when 'in contact:' smell ranges over, at the utmost, a mile or two, and the greatest distance which sound is ever known to have traveled (when Krakatoa exploded in 1883) is but a few thousand miles—a mere fraction of the earth's girdle."—Prof. H. H. Turner of Oxford.

Note 2 (page 27)

Huyghens and Leibniz both objected to Newton's inverse square law because it postulated "action at a distance,"—for example, the attractive force of the sun and the earth. This desire for "continuity" in physical laws led to the supposition of an "ether." We may here anticipate and state that the reason which prompted Huyghens to object to Newton's law led Einstein in our own day to raise objections to the "ether" theory. "In the formulation of physical laws, only those things were to be regarded as being in causal connection which were capable of being actually observed." And the "ether" has not been "actually observed."

The idea of "continuity" implies distances between adjacent points that are infinitesimal in extent; hence the idea of "continuity" comes in direct opposition with the finite distances of Newton.

The statement relating to causal connection—the refusal to accept an "ether" as an absolute base of reference—leads to the principle of the relativity of motion.

NOTE 3 (page 30)

Sir Oliver Lodge goes to the extreme of pinning his faith in the reality of this ether rather than in that of matter. Witness the following statement he made recently before a New York audience:

"To my mind the ether of space is a substantial reality with extraordinarily perfect properties, with an immense amount of energy stored up in it, with a constitution which we must discover, but a substantial reality far more impressive than that of matter. Empty space, as we call it, is full of ether, but it makes no appeal to our senses. The appearance is as if it were nothing. It is the most important thing in the material universe. I believe that matter is a modification of ether, a very porous substance, a thing more analogous to a cobweb or the Milky Way or something very slight and unsubstantial, as compared to ether."

And again:

"The properties of ether seem to be perfect. Matter is less so; it has friction and elasticity. No imperfection has been discovered in the ether space. It doesn't wear out; there is no dissipation of energy; there is no friction. Ether is material, yet it is not matter; both are substantial realities in physics, but it is the ether of space that holds things together and acts as a cement. My business is to call attention to the whole world of etherealness of things, and I have made it a subject of thirty years' study, but we must admit that there is no getting hold of ether except indirectly."

"I consider the ether of space," says Lodge, in conclusion, "the one substantial thing in the universe." And Lodge is certainly entitled to his opinion.

NOTE 4 (page 51)

For the benefit of those readers who wish to gain a deeper insight into the relativity principle, we shall here discuss it very briefly.

Newton and Galileo had developed a relativity principle in mechanics which may be stated as follows: If one system of reference is in uniform rectilinear motion with respect to another system of reference, then whatever physical laws are deduced from the first system hold true for the second system. The two systems are *equivalent*. If the two systems be represented by xyz and

,

and if they move with the velocity of v along the x-axis with respect to one another, then the two systems are mathematically related thus:

$$x' = x - vt, y' = y, z' = z, t' = t,$$

(1)

and this immediately provides us with a means of transforming the laws of one system to those of another.

With the development of electrodynamics (which we may call electricity in motion) difficulties arose which equations in mechanics of type (1) could no longer solve. These difficulties merely increased when Maxwell showed that light must be regarded as an electromagnetic phenomenon. For suppose we wish to investigate the motion of a source of light (which may be the equivalent of the motion of the earth with reference to the sun) with respect to the velocity of the light it emits—a typical example of the study of *moving systems*— how are we to coordinate the electrodynamical and mechanical elements? Or, again, suppose we wish to investigate the velocity of electrons shot out from radium with a speed comparable to that of light, how are we to coordinate the two branches in tracing the course of these negative particles of electricity?

It was difficulties such as these that led to the Lorentz-Einstein modifications of the Newton-Galileo relativity equations (1). The Lorentz-Einstein equations are expressed in the form:

$$x' = \frac{x - vt}{\sqrt{1 - \frac{v^2}{c}}}, y' = y, z' = z, t' = \frac{t - \frac{v}{c^2} \cdot x}{\sqrt{1 - \frac{v^2}{c^2}}},$$

(2)

c denoting the velocity of light *in vacuo* (which, according to all observations, is the same, irrespective of the observer's state of motion). Here, you see, electrodynamical systems (light and therefore "ray" velocities such as those due to electrons) are brought into play.

This gives us Einstein's *special theory of relativity*. From it Einstein deduced some startling conceptions of time and space.

NOTE 5 (page 55)

The velocity (v) of an object in one system will have a different velocity (v') if referred to another system in uniform motion relative to the first. It had been supposed that only a "something" endowed with *infinite* velocity would show the *same* velocity in *all* systems, irrespective of the motions of the latter. Michelson and Morley's results actually point to the velocity of light as showing the properties of the imaginary "infinite velocity." The velocity of light possesses universal significance; and this is the basis for much of Einstein's earlier work.

NOTE 6 (page 56)

"Euclid assumes that parallel lines never meet, which they cannot do of course if they be defined as equidistant. But are there such lines? And if not, why not assume that all lines drawn through a point outside a given line will eventually intersect it? Such an assumption leads to a geometry in which all lines are conceived as being drawn on the surface of a sphere or an ellipse, and in it the three angles of a triangle are never quite equal to two right angles, nor the circumference of a circle quite π times its diameter. *But that is precisely what the contraction effect due to motion requires.*"

(Dr. Walker)

Note 7 (page 57)

Einstein had become tired of assumptions. He had no particular objection to the "ether" theory beyond the fact that this "ether" did not come within the range of our senses; it could not be "observed." "The consistent fulfilment of the two postulates—'action by contact' and causal relationship between only such things as lie within the realm of observation [see Note 2] combined together is, I believe, the mainspring of Einstein's method of investigation...." (Prof. Freundlich).

Note 8 (page 59)

That the conception of the "simultaneity" of events is devoid of meaning can be deduced from equation (2) [see Note 4]. We owe the proof to Einstein. "It is possible to select a suitable time-coordinate in such a way that a time-measurement enters into physical laws in exactly the same manner as regards its significance as a space measurement (that is, they are fully equivalent symbolically), and has likewise a definite coordinate direction.... It never occurred to anyone that the use of a light-signal as a means of connection between the moving-body and the observer, which is necessary in practice in order to determine *simultaneity*, might affect the final result, *i.e.*, of time measurements in different systems." (Freundlich). But that is just what Einstein shows, because time-measurements are based on "simultaneity of events," and this, as pointed out above, is devoid of meaning.

Had the older masters the occasion to study enormous velocities, such as the velocity of light, rather than relatively small ones—and even the velocity of the earth around the sun is small as compared to the velocity of light—discrepancies between theory and experiment would have become apparent.

NOTE 9 (page 67)

How the *special* theory of relativity (see Note 4) led to the *general* theory of relativity (which included gravitation) may now be briefly traced.

When we speak of electrons, or negative particles of electricity, in motion, we are speaking of *energy* in motion. Now these electrons when in motion exhibit properties that are very similar to matter in motion. Whatever deviations

65

there are are due to the enormous velocity of these electrons, and this velocity, as has already been pointed out, is comparable to that of light; whereas before the advent of the electron, the velocity of no particles comparable to that of light had ever been measured.

According to present views "all inertia of matter consists only of the inertia of the latent energy in it; ... everything that we know of the inertia of energy holds without exception for the inertia of *matter*."

Now it is on the assumption that inertial mass and gravitational "pull" are equivalent that the mass of a body is determined by its *weight*. What is true of matter should be true of energy.

The special theory of relativity, however, takes into account only inertia ("inertial *mass*") but not gravitation (gravitational pull or *weight*) of energy. When a body absorbs energy equation 2 (see Note 4) will record a gain in inertia but not in weight—which is contrary to one of the fundamental facts in mechanics.

This means that a more *general* theory of relativity is required to include gravitational phenomena. Hence Einstein's *General Theory of Relativity*. Hence the approach to a new theory of gravitation. Hence "the setting up of a differential equation which comprises the motion of a body under the influence of both *inertia and gravity*, and which symbolically expresses the relativity of motions.... The differential law must always preserve the same form, irrespective of the system of coordinates to which it is referred, so that no system of coordinates enjoys a preference to any other." (For the general form of the equation and for an excellent discussion of its significance, see Freundlich's monograph, pages 27–33.)

TIME, SPACE, AND GRAVITATION[1]

BY
PROF. ALBERT EINSTEIN

There are several kinds of theory in physics. Most of them are constructive. These attempt to build a picture of complex phenomena out of some relatively simple proposition. The kinetic theory of gases, for instance, attempts to refer to molecular movement the mechanical thermal, and diffusional properties of gases. When we say that we understand a group of natural phenomena, we mean that we have found a constructive theory which embraces them.

Theories of Principle.—But in addition to this most weighty group of theories, there is another group consisting of what I call theories of principle. These employ the analytic, not the synthetic method. Their starting-point and foundation are not hypothetical constituents, but empirically observed general properties of phenomena, principles from which mathematical formulæ are deduced of such a kind that they apply to every case which presents itself. Thermodynamics, for instance, starting from the fact that perpetual motion never occurs in ordinary experience, attempts to deduce from this, by analytic processes, a theory which will apply in every case. The merit of constructive theories is their comprehensiveness, adaptability, and clarity, that of the theories of principle, their logical perfection, and the security of their foundation.

The theory of relativity is a theory of principle. To understand it, the principles on which it rests must be grasped. But before stating these it is necessary to point out that the theory of relativity is like a house with two separate stories, the special relativity theory and the general theory of relativity.

Since the time of the ancient Greeks it has been well known that in describing the motion of a body we must refer to another body. The motion of a railway train is described with reference to the ground, of a planet with reference to the total assemblage of visible fixed stars. In physics the bodies to which motions are spatially referred are termed systems of coordinates. The laws of mechanics of Galileo and Newton can be formulated only by using a system of coordinates.

The state of motion of a system of coordinates can not be chosen arbitrarily if the laws of mechanics are to hold good (it must be free from twisting and from acceleration). The system of coordinates employed in mechanics is called an inertia-system. The state of motion of an inertia-system, so far as mechanics are concerned, is not restricted by nature to one condition. The condition in the following proposition suffices; a system of coordinates moving in the same direction and at the same rate as a system of inertia is itself a system of inertia. The special relativity theory is therefore the application of the following proposition to any natural process: "Every law of nature which holds good with respect to a coordinate system K must also hold good for any other system K' provided that K and K' are in uniform movement of translation."

The second principle on which the special relativity theory rests is that of the constancy of the velocity of light in a vacuum. Light in a vacuum has a definite and constant velocity, independent of the velocity of its source. Physicists owe their confidence in this proposition to the Maxwell-Lorentz theory of electrodynamics.

The two principles which I have mentioned have received strong experimental confirmation, but do not seem to be logically compatible. The special relativity theory achieved their logical reconciliation by making a change in kinematics, that is to say, in the doctrine of the physical laws of space and time. It became evident that a statement of the coincidence of two events could have a meaning only in connection with a system of coordinates, that the mass of bodies and the rate of movement of clocks must depend on their state of motion with regard to the coordinates.

The Older Physics.—But the older physics, including the laws of motion of Galileo and Newton, clashed with the relativistic kinematics that I have indicated. The latter gave origin to certain generalized mathematical conditions with which the laws of nature would have to conform if the two fundamental principles were compatible. Physics had to be modified. The most notable change was a new law of motion for (very rapidly) moving mass-points, and this soon came to be verified in the case of electrically-laden particles. The most important result of the special relativity system concerned the inert mass of a material system. It became evident that the inertia of such a system must depend on its energy-content, so that we were driven to the conception that inert mass was nothing else than latent energy. The doctrine of the conservation of mass lost its independence and became merged in the doctrine of conservation of energy.

The special relativity theory which was simply a systematic extension of the electrodynamics of Maxwell and Lorentz, had consequences which reached beyond itself. Must the independence of physical laws with regard to a system of coordinates be limited to systems of coordinates in uniform movement of translation with regard to one another? What has nature to do with the coordinate systems that we propose and with their motions? Although it may be necessary for our descriptions of nature to employ systems of coordinates that we have selected arbitrarily, the choice should not be limited in any way so far as their state of motion is concerned. (General theory of relativity.) The application of this general theory of relativity was found to be in conflict with a well-known experiment, according to which it appeared that the weight and the inertia of a body depended on the same constants (identity of inert and heavy masses). Consider the case of a system of coordinates which is conceived as being in stable rotation relative to a system of inertia in the Newtonian sense. The forces which, relatively to this system, are centrifugal must, in the Newtonian sense, be attributed to inertia. But these centrifugal forces are, like gravitation, proportional to the mass of the bodies. Is it not, then, possible to regard the system of coordinates as at rest, and the centrifugal forces as gravitational? The interpretation seemed obvious, but classical mechanics forbade it.

This slight sketch indicates how a generalized theory of relativity must include the laws of gravitation, and actual pursuit of the conception has justified the hope. But the way was harder than was expected, because it contradicted Euclidian geometry. In other words, the laws according to which material bodies are arranged in space do not exactly agree with the laws of space prescribed by the Euclidian geometry of solids. This is what is meant by the phrase "a warp in space." The fundamental concepts "straight," "plane," etc., accordingly lose their exact meaning in physics.

In the generalized theory of relativity, the doctrine of space and time, kinematics, is no longer one of the absolute foundations of general physics. The geometrical states of bodies and the rates of clocks depend in the first place on their gravitational fields, which again are produced by the material system concerned.

Thus the new theory of gravitation diverges widely from that of Newton with respect to its basal principle. But in practical application the two agree so closely that it has been difficult to find cases in which the actual differences could be subjected to observation. As yet only the following have been suggested:

1. The distortion of the oval orbits of planets round the sun (confirmed in the case of the planet Mercury).

2. The deviation of light-rays in a gravitational field (confirmed by the English Solar Eclipse expedition).

3. The shifting of spectral lines towards the red end of the spectrum in the case of light coming to us from stars of appreciable mass (not yet confirmed).

The great attraction of the theory is its logical consistency. If any deduction from it should prove untenable, it must be given up. A modification of it seems impossible without destruction of the whole.

No one must think that Newton's great creation can be overthrown in any real sense by this or by any other theory. His clear and wide ideas will for ever retain their significance as the foundation on which our modern conceptions of physics have been built.

1

Republished by permission from "Science."

EINSTEIN'S LAW OF GRAVITATION[1]

BY
PROF. *J. S. AMES*
Johns Hopkins University

... In the treatment of Maxwell's equations of the electromagnetic field, several investigators realized the importance of deducing the form of the equations when applied to a system moving with a uniform velocity. One object of such an investigation would be to determine such a set of transformation formulæ as would leave the mathematical form of the equations unaltered. The necessary relations between the new space-coordinates, those applying to the moving system, and the original set were of course obvious; and elementary methods led to the deduction of a new variable which should replace the time coordinate. This step was taken by Lorentz and also, I believe, by Larmor and by Voigt. The mathematical deductions and applications in the hands of these men were extremely beautiful, and are probably well known to you all.

Lorentz' paper on this subject appeared in the Proceedings of the Amsterdam Academy in 1904. In the following year there was published in the *Annalen der Physik* a paper by Einstein, written without any knowledge of the work of Lorentz, in which he arrived at the same transformation equations as did the latter, but with an entirely different and fundamentally new interpretation. Einstein called attention in his paper to the lack of definiteness in the concepts of time and space, as ordinarily stated and used. He analyzed clearly the definitions and postulates which were necessary before one could speak with exactness of a length or of an interval of time. He disposed forever of the propriety of speaking of the "true" length of a rod or of the "true" duration of time, showing, in fact, that the numerical values which we attach to lengths or intervals of time depend upon the definitions and postulates which we adopt. The words "absolute" space or time intervals are devoid of meaning. As an illustration of what is meant Einstein discussed two possible ways of measuring the length of a rod when it is moving in the direction of its own length with a uniform velocity, that is, after having adopted a scale of length, two ways of assigning a number to the length of the rod concerned. One method is to imagine the observer moving with the rod, applying along its length the measuring scale, and reading off the positions of the ends of the rod. Another method would be to have two observers at rest on the body with reference to which the rod has the uniform velocity, so stationed along the line of motion of the rod that as the rod moves past them they can note simultaneously on a stationary measuring scale the positions of the two ends of the rod. Einstein showed that, accepting two postulates which need no defense at this time, the two methods of measurements would lead to different numerical values, and, further, that the divergence of the two results would increase as the velocity of the rod was increased. In assigning a number, therefore, to the length of a moving rod, one must make a choice of the method to be used in measuring it. Obviously the preferable method is to agree that the observer shall move with the rod, carrying his measuring instrument with him. This disposes of the problem of measuring space relations. The observed fact that, if we measure the length of the rod on different days, or when the rod is lying in different positions, we always obtain the same value offers no information concerning the "real" length of the rod. It may have changed, or it may not. It must always be remembered that measurement of the length of a rod is simply a process of comparison between it and an arbitrary standard, *e.g.*, a meter-rod or yard-stick. In regard to the problem of assigning numbers to intervals of time, it must be borne in mind that, strictly speaking, we do not "measure" such intervals, *i.e.*, that we do not select a unit interval of time and find how many times it is contained in the interval in question. (Similarly, we do not "measure" the pitch of a sound or the temperature of a room.) Our practical instruments for assigning numbers to time-intervals depend in the main upon our agreeing to believe that a pendulum swings in a perfectly uniform manner, each vibration taking the same time as the next one. Of course we cannot *prove* that this is true, it is, strictly speaking, a definition of what we mean by equal intervals of time; and it is not a particularly good definition at that. Its limitations are sufficiently obvious. The best way to proceed is to consider the concept of uniform velocity, and then, using the idea of some entity having such a uniform velocity, to define equal intervals of time as such intervals as are required for the entity to traverse equal lengths. These last we have already defined. What is required in addition is to adopt some moving entity as giving our definition of uniform velocity. Considering our known universe it is self-evident that we should choose in our definition of uniform velocity the velocity of light, since this selection could be made by an observer anywhere in our universe. Having agreed then to illustrate by the words "uniform velocity" that of light, our definition of equal intervals of time is complete. This implies, of course, that there is no uncertainty on our part as to the fact that the velocity of light always has the same value at any one point in the universe to any observer, quite regardless of the source of light. In other words, the postulate that this is true underlies our definition. Following this method Einstein developed a system of measuring both space and time intervals. As a matter of fact his system is identically that which we use in daily life with reference to events here on the earth. He further showed that if a man were to measure the length of a rod, for instance, on the earth and then were able to carry the rod and his measuring apparatus to Mars, the sun, or to Arcturus he would obtain the same numerical value for the length in all places and at all times. This doesn't mean that any statement is implied as to whether the length of the rod has remained unchanged or not; such words do not have any meaning—remember that we can not speak of true length. It is thus clear that an observer living on the earth would have a definite system of units in terms of which to express space and time intervals, *i.e.*, he would have a definite system of space coordinates (x, y, z) and a definite time coordinate (t); and similarly an observer living on Mars would have his system of coordinates (x', y', z', t'). Provided that one observer has a definite uniform velocity with reference to the other, it is a comparatively simple matter to deduce the mathematical relations between the two sets of coordinates. When Einstein did this, he arrived at the same transformation formulæ as those used by Lorentz in his development of Maxwell's equations. The latter had shown that, using these formulæ, the form of the laws for all electromagnetic phenomena maintained the same form; so Einstein's method proves that using his system of measurement an observer, anywhere in the universe, would as the result of his own investigation of electromagnetic phenomena arrive at the same mathematical statement of them as any other observer, provided only that the relative-velocity of the two observers was uniform.

Einstein discussed many other most important questions at this time; but it is not necessary to refer to them in connection with the present subject. So far as this is concerned, the next important step to note is that taken in the famous address of Minkowski, in 1908, on the subject of "Space and Time." It would be difficult to overstate the importance of the concepts advanced by Minkowski. They marked the beginning of a new period in the philosophy of physics. I shall not attempt to explain his ideas in detail, but shall confine myself to a few general statements. His point of view and his line of development of the theme are absolutely different from those of Lorentz or of Einstein; but in the end he makes use of the same transformation formulæ. His great contribution consists in giving us a new geometrical picture of their meaning. It is scarcely fair to call Minkowski's development a picture; for to us a picture can never have more than three dimensions, our senses limit us; while his picture calls for perception of four dimensions. It is this fact that renders any even semi-popular discussion of Minkowski's work so impossible. We can all see that for us to describe any event a knowledge of four coordinates is necessary, three for the space specification and one for the time. A complete picture could be given then by a point in four dimensions. All four coordinates are necessary: we never observe an event except at a certain time, and we never observe an instant of time except with reference to space. Discussing the laws of electromagnetic phenomena, Minkowski showed how in a space of four dimensions, by a suitable definition of axes, the mathematical transformation of Lorentz and Einstein could be described by a rotation of the set of axes. We are all accustomed to a rotation of our ordinary cartesian set of axes describing the position of a point. We ordinarily choose our axes at any location on the earth as follows: one vertical, one east and west, one north and south. So if we move from any one laboratory to another, we change our axes; they are always orthogonal, but in moving from place to place there is a rotation. Similarly, Minkowski showed that if we choose four orthogonal axes at any point on the earth, according to his method, to represent a space-time point using the method of measuring space and time intervals as outlined by Einstein; and, if an observer on Arcturus used a similar set of axes and the method of measurement which he naturally would, the set of axes of the latter could be obtained from those of the observer on the earth by a pure rotation (and naturally a transfer of the origin). This is a beautiful geometrical result. To complete my statement of the method, I must add that instead of using as his fourth axis one along which numerical values of time are laid off, Minkowski defined his fourth coordinate as the product of time and the imaginary constant, the square root of minus one. This introduction of imaginary quantities might be expected, possibly, to introduce difficulties; but, in reality, it is the very essence of the simplicity of the geometrical description just given of the rotation of the sets of axes. It thus appears that different observers situated at different points in the universe would each have their own set of axes, all different, yet all connected by the fact that any one can be rotated so as to coincide with any other. This means that there is no one direction in the four-dimensional space that corresponds to time for all observers. Just as with reference to the earth there is no direction which can be called vertical for all observers living on the earth. In the sense of an *absolute* meaning the words "up and down," "before and after," "sooner or later," are entirely meaningless.

This concept of Minkowski's may be made clearer, perhaps, by the following process of thought. If we take a section through our three-dimensional space, we have a plane, *i.e.*, a two-dimensional space. Similarly, if a section is made through a four-dimensional space, one of three dimensions is obtained. Thus, for an observer on the earth a definite section of Minkowski's four-dimensional space will give us our ordinary three-dimensional one; so that this section will, as it were, break up Minkowski's space into our space and give us our ordinary time. Similarly, a different section would have to be used to the observer on Arcturus; but by a suitable selection he would get his own familiar three-dimensional space and his own time. Thus the space defined by Minkowski is completely isotropic in reference to measured lengths and times, there is absolutely no difference between any two directions in an absolute sense; for any particular observer, of course, a particular section will cause the space to fall apart so as to suit his habits of measurement; any section, however, taken at random will do the same thing for some observer somewhere. From another point of view, that of Lorentz and Einstein, it is obvious that, since this four-dimensional space is isotropic, the expression of the laws of electromagnetic phenomena take identical mathematical forms when expressed by any observer.

The question of course must be raised as to what can be said in regard to phenomena which so far as we know do not have an electromagnetic origin. In particular what can be done with respect to gravitational phenomena? Before, however, showing how this problem was attacked by Einstein; and the fact that the subject of my address is Einstein's work on gravitation shows that ultimately I shall explain this, I must emphasize another feature of Minkowski's geometry. To describe the space-time characteristics of any event a point, defined by its four coordinates, is sufficient; so, if one observes the life-history of any entity, *e.g.*, a particle of matter, a light-wave, etc., he observes a sequence of points in the space-time continuum; that is, the life-history of any entity is described fully by a line in this space. Such a line was called by Minkowski a "world-line." Further, from a different point of view, all of our observations of nature are in reality observations of coincidences, *e.g.*, if one reads a thermometer, what he does is to note the coincidence of the end of the column of mercury with a certain scale division on the thermometer tube. In other words, thinking of the world-line of the end of the mercury column and the world-line of the scale division, what we have observed was the intersection or crossing of these lines. In a similar manner any observation may be analyzed; and remembering that light rays, a point on the retina of the eye, etc., all have their world-lines, it will be recognized that it is a perfectly accurate statement to say that every observation is the perception of the intersection of world-lines. Further, since all we know of a world-line is the result of observations, it is evident that we do not know a world-line as a continuous series of points, but simply as a series of discontinuous points, each point being where the particular world-line in question is crossed by another world-line.

It is clear, moreover, that for the description of a world-line we are not limited to the particular set of four orthogonal axes adopted by Minkowski. We can choose any set of four-dimensional axes we wish. It is further evident that the mathematical expression for the coincidence of two points is absolutely independent of our selection of reference axes. If we change our axes, we will change the coordinates of both points simultaneously, so that the question of axes ceases to be of interest. But our so-called laws of nature are nothing but descriptions in mathematical language of our observations; we observe only coincidences; a sequence of coincidences when put in mathematical terms takes a form which is independent of the selection of reference axes; therefore the mathematical expression of our laws of nature, of every character, must be such that their form does not change if we make a transformation of axes. This is a simple but far-reaching deduction.

There is a geometrical method of picturing the effect of a change of axes of reference, *i.e.*, of a mathematical transformation. To a man in a railway coach the path of a drop of water does not appear vertical, *i.e.*, it is not parallel to the edge of the window; still less so does it appear vertical to a man performing manœuvres in an airplane. This means that whereas with reference to axes fixed to the earth the path of the drop is vertical; with reference to other axes, the path is not. Or, stating the conclusion in general language, changing the axes of reference (or effecting a mathematical transformation) in general changes the shape of any line. If one imagines the line forming a part of the space, it is evident that if the space is deformed by compression or expansion the shape of the line is changed, and if sufficient care is taken it is clearly possible, by deforming the space, to make the line take any shape desired, or better stated, any shape specified by the previous change of axes. It is thus possible to picture a mathematical transformation as a deformation of space. Thus I can draw a line on a sheet of paper or of rubber and by bending and stretching the sheet, I can make the line assume a great variety of shapes; each of these new shapes is a picture of a suitable transformation.

Now, consider world-lines in our four-dimensional space. The complete record of all our knowledge is a series of sequences of intersections of such lines. By analogy I can draw in ordinary space a great number of intersecting lines on a sheet of rubber; I can then bend and deform the sheet to please myself; by so doing I do not introduce any new intersections nor do I alter in the least the sequence of intersections. So in the space of our world-lines, the space may be deformed in any imaginable manner without introducing any new intersections or changing the sequence of the existing intersections. It is this sequence which gives us the mathematical expression of our so-called experimental laws; a deformation of our space is equivalent mathematically to a transformation of axes, consequently we see why it is that the form of our laws must be the same when referred to any and all sets of axes, that is, must remain unaltered by any mathematical transformation.

Now, at last we come to gravitation. We can not imagine any world-line simpler than that of a particle of matter left to itself; we shall therefore call it a "straight" line. Our experience is that two particles of matter attract one another. Expressed in terms of world-lines, this means that, if the world-lines of two isolated particles come near each other, the lines, instead of being straight, will be deflected or bent in towards each other. The world-line of any one particle is therefore deformed; and we have just seen that a deformation is the equivalent of a mathematical transformation. In other words, for any one particle it is possible to replace the effect of a gravitational field at any instant by a mathematical transformation of axes. The statement that this is always possible for any particle at any instant is Einstein's famous "Principle of Equivalence."

Let us rest for a moment, while I call attention to a most interesting coincidence, not to be thought of as an intersection of world-lines. It is said that Newton's thoughts were directed to the observation of gravitational phenomena by an apple falling on his head; from this striking event he passed by natural steps to a consideration of the universality of gravitation. Einstein in describing his mental process in the evolution of his law of gravitation says that his attention was called to a new point of view by discussing his experiences with a man whose fall from a high building he had just witnessed. The man fortunately suffered no serious injuries and assured Einstein that in the course of his fall he had not been conscious in the least of any pull downward on his body. In mathematical language, with reference to axes moving with the man the force of gravity had disappeared. This is a case where by the transfer of the axes from the earth itself to the man, the force of the gravitational field is annulled. The converse change of axes from the falling man to a point on the earth could be considered as introducing the force of gravity into the equations of motion. Another illustration of the introduction into our equations of a force by a means of a change of axes is furnished by the ordinary treatment of a body in uniform rotation about an axis. For instance, in the case of a so-called conical pendulum, that is, the motion of a bob suspended from a fixed point by string, which is so set in motion that the bob describes a horizontal circle and the string therefore describes a circular cone, if we transfer our axes from the earth and have them rotate around the vertical line through the fixed point with the same angular velocity as the bob, it is necessary to introduce into our equations of motion a fictitious "force" called the centrifugal force. No one ever thinks of this force other than as a mathematical quantity introduced into the equations for the sake of simplicity of treatment; no physical meaning is attached to it. Why should there be to any other so-called "force," which like centrifugal force, is independent of the nature of the matter? Again, here on the earth our sensation of weight is interpreted mathematically by combining expressions for centrifugal force and gravity; we have no distinct sensation for either separately. Why then is there any difference in the essence of the two? Why not consider them both as brought into our equations by the agency of mathematical transformations? This is Einstein's point of view.

Granting, then, the principle of equivalence, we can so choose axes at any point at any instant that the gravitational field will disappear; these axes are therefore of what Eddington calls the "Galilean" type, the simplest possible. Consider, that is, an observer in a box, or compartment, which is falling with the acceleration of the gravitational field at that point. He would not be conscious of the field. If there were a projectile fired off in this compartment, the observer would describe its path as being straight. In this space the infinitesimal interval between two space-time points would then be given by the formula

$$ds^2 = dx_1^2 + dx2_2 + dx_3^2 + dx2_4,$$

where ds is the interval and

$$x_1, x_2, x_3, x_4$$

are coordinates. If we make a mathematical transformation, *i.e.*, use another set of axes, this interval would obviously take the form

$$ds^2 = g_{11}dx_{33}^2 + g_{22}dx_2^2 + g_{33}dx_3^2 + g_{44}dx2_4 + 2g_{12}dx_1dx_2 + \text{etc.},$$

where

$$x_1, x_2, x_3 \text{ and } x_4$$

are now coordinates referring to the new axes. This relation involves ten coefficients, the coefficients defining the transformation.

But of course a certain dynamical value is also attached to the *g*'s, because by the transfer of our axes from the Galilean type we have made a change which is equivalent to the introduction of a gravitational field; and the *g*'s must specify the field. That is, these *g*'s are the expressions of our experiences, and hence their values can not depend upon the use of any special axes; the values must be the same for all selections. In other words, whatever function of the coordinates any one *g* is for one set of axes, if other axes are chosen, this *g* must still be the same function of the new coordinates. There are ten g's defined by differential equations; so we have ten covariant equations. Einstein showed how these g's could be regarded as generalized potentials of the field. Our own experiments and observations upon gravitation have given us a certain knowledge concerning its potential; that is, we know a value for it which must be so near the truth that we can properly call it at least a first approximation. Or, stated differently, if Einstein succeeds in deducing the rigid value for the gravitational potential in any field, it must degenerate to the Newtonian value for the great majority of cases with which we have actual experience. Einstein's method, then, was to investigate the functions (or equations) which would satisfy the mathematical conditions just described. A transformation from the axes used by the observer in the following box may be made so as to introduce into the equations the gravitational field recognized by an observer on the earth near the box; but this, obviously, would not be the general gravitational field, because the field changes as one moves over the surface of the earth. A solution found, therefore, as just indicated, would not be the one sought for the general field; and another must be found which is less stringent than the former but reduces to it as a special case. He found himself at liberty to make a selection from among several possibilities, and for several reasons chose the simplest solution. He then tested this decision by seeing if his formulæ would degenerate to Newton's law for the limiting case of velocities small when compared with that of light, because this condition is satisfied in those cases to which Newton's law applies. His formulæ satisfied this test, and he therefore was able to announce a "law of gravitation," of which Newton's was a special form for a simple case.

To the ordinary scholar the difficulties surmounted by Einstein in his investigations appear stupendous. It is not improbable that the statement which he is alleged to have made to his editor, that only ten men in the world could understand his treatment of the subject, is true. I am fully prepared to believe it, and wish to add that I certainly am not one of the ten. But I can also say that, after a careful and serious study of his papers, I feel confident that there is nothing in them which I can not understand, given the time to become familiar with the special mathematical processes used. The more I work over Einstein's papers, the more impressed I am, not simply by his genius in viewing the problem, but also by his great technical skill.

Following the path outlined, Einstein, as just said, arrived at certain mathematical laws for a gravitational field, laws which reduced to Newton's form in most cases where observations are possible, but which led to different conclusions in a few cases, knowledge concerning which we might obtain by careful observations. I shall mention a few deductions from Einstein's formulæ.

1. If a heavy particle is put at the center of a circle, and, if the length of the circumference and the length of the diameter are measured, it will be found that their ratio is not π (3.14159). In other words the geometrical properties of space in such a gravitational field are not those discussed by Euclid; the space is, then, non-Euclidean. There is no way by which this deduction can be verified, the difference between the predicted ratio and π is too minute for us to hope to make our measurements with sufficient exactness to determine the difference.

2. All the lines in the solar spectrum should with reference to lines obtained by terrestrial sources be displaced slightly towards longer wave-lengths. The amount of displacement predicted for lines in the blue end of the spectrum is about one-hundredth of an Angstrom unit, a quantity well within experimental limits. Unfortunately, as far as the testing of this prediction is concerned, there are several physical causes which are also operating to cause displacement of the spectrum-lines; and so at present a decision can not be rendered as to the verification. St. John and other workers at the Mount Wilson Observatory have the question under investigation.

3. According to Newton's law an isolated planet in its motion around a central sun would describe, period after period, the same elliptical orbit; whereas Einstein's laws lead to the prediction that the successive orbits traversed would not be identically the same. Each revolution would start the planet off on an orbit very approximately elliptical, but with the major axis of the ellipse rotated slightly in the plane of the orbit. When calculations were made for the various planets in our solar system, it was found that the only one which was of interest from the standpoint of verification of Einstein's formulæ was Mercury. It has been known for a long time that there was actually such a change as just described in the orbit of Mercury, amounting to 574" of arc per century; and it has been shown that of this a rotation of 532" was due to the direct action of other planets, thus leaving an unexplained rotation of 42" per century. Einstein's formulæ predicted a rotation of 43", a striking agreement.

4. In accordance with Einstein's formulæ a ray of light passing close to a heavy piece of matter, the sun, for instance, should experience a sensible deflection in towards the sun.

This might be expected from "general" consideration of energy in motion; energy and mass are generally considered to be identical in the sense that an amount of energy E has the

$$E1c^2$$

mass where c is the velocity of light; and consequently a ray of light might fall within the province of gravitation and the amount of deflection to be expected could be calculated by the ordinary formula for gravitation. Another point of view is to consider again the observer inside the compartment falling with the acceleration of the gravitational field. To him the path of a projectile and a ray of light would both appear straight; so that, if the projectile had a velocity equal to that of light, it and the light wave would travel side by side. To an observer outside the compartment, *e.g.*, to one on the earth, both would then appear to have the same deflection owing to the sun. But how much would the path of the projectile be bent? What would be the shape of its parabola? One might apply Newton's law; but, according to Einstein's formulæ, Newton's law should be used only for small velocities. In the case of a ray passing close to the sun it was decided that according to Einstein's formula there should be a deflection of $1''.75$ whereas Newton's law of gravitation predicted half this amount. Careful plans were made by various astronomers, to investigate this question at the solar eclipse last May, and the result announced by Dyson, Eddington and

Crommelin, the leaders of astronomy in England, was that there was a deflection of $1''.9$. Of course the detection of such a minute deflection was an extraordinarily difficult matter, so many corrections had to be applied to the original observations; but the names of the men who record the conclusions are such as to inspire confidence. Certainly any effect of refraction seems to be excluded.

It is thus seen that the formulæ deduced by Einstein have been confirmed in a variety of ways and in a most brilliant manner. In connection with these formulæ one question must arise in the minds of everyone; by what process, where in the course of the mathematical development, does the idea of mass reveal itself? It was not in the equations at the beginning and yet here it is at the end. How does it appear? As a matter of fact it is first seen as a constant of integration in the discussion of the problem of the gravitational field due to a single particle; and the identity of this constant with mass is proved when one compares Einstein's formulæ with Newton's law which is simply its degenerated form. This mass, though, is the mass of which we become aware through our experiences with weight; and Einstein proceeded to prove that this quantity which entered as a constant of integration in his ideally simple problem also obeyed the laws of conservation of mass and conservation of momentum when he investigated the problems of two and more particles. Therefore Einstein deduced from his study of gravitational fields the well-known properties of matter which form the basis of theoretical mechanics. A further logical consequence of Einstein's development is to show that energy has mass, a concept with which every one nowadays is familiar.

The description of Einstein's method which I have given so far is simply the story of one success after another; and it is certainly fair to ask if we have at last reached finality in our investigation of nature, if we have attained to truth. Are there no outstanding difficulties? Is there no possibility of error? Certainly, not until all the predictions made from Einstein's formulæ have been investigated can much be said; and further, it must be seen whether any other lines of argument will lead to the same conclusions. But without waiting for all this there is at least one difficulty which is apparent at this time. We have discussed the laws of nature as independent in their form of reference axes, a concept which appeals strongly to our philosophy; yet it is not at all clear, at first sight, that we can be justified in our belief. We can not imagine any way by which we can become conscious of the translation of the earth in space; but by means of gyroscopes we can learn a great deal about its rotation on its axis. We could locate the positions of its two poles, and by watching a Foucault pendulum or a gyroscope we can obtain a number which we interpret as the angular velocity of rotation of axes fixed in the earth; angular velocity with reference to what? Where is the fundamental set of axes? This is a real difficulty. It can be surmounted in several ways. Einstein himself has outlined a method which in the end amounts to assuming the existence on the confines of space of vast quantities of matter, a proposition which is not attractive. deSitter has suggested a peculiar quality of the space to which we refer our space-time coordinates. The consequences of this are most interesting, but no decision can as yet be made as to the justification of the hypothesis. In any case we can say that the difficulty raised is not one that destroys the real value of Einstein's work.

In conclusion I wish to emphasize the fact, which should be obvious, that Einstein has not attempted any explanation of gravitation; he has been occupied with the deduction of its laws. These laws, together with those of electromagnetic phenomena, comprise our store of knowledge. There is not the slightest indication of a mechanism, meaning by that a picture in terms of our senses. In fact what we have learned has been to realize that our desire to use such mechanisms is futile.

---

[1]

Presidential address delivered at the St. Louis meeting of the Physical Society, December 30, 1919. Republished by permission from "Science." ↑

THE DEFLECTION OF LIGHT BY GRAVITATION AND THE EINSTEIN THEORY OF RELATIVITY.1

SIR FRANK DYSON
the Astronomer Royal

The purpose of the expedition was to determine whether any displacement is caused to a ray of light by the gravitational field of the sun, and if so, the amount of the displacement. Einstein's theory predicted a displacement varying inversely as the distance of the ray from the sun's center, amounting to 1".75 for a star seen just grazing the sun....

A study of the conditions of the 1919 eclipse showed that the sun would be very favorably placed among a group of bright stars—in fact, it would be in the most favorable possible position. A study of the conditions at various points on the path of the eclipse, in which Mr. Hinks helped us, pointed to Sobral, in Brazil, and Principe, an island off the west coast of Africa, as the most favorable stations....

The Greenwich party, Dr. Crommelin and Mr. Davidson, reached Brazil in ample time to prepare for the eclipse, and the usual preliminary focusing by photographing stellar fields was carried out. The day of the eclipse opened cloudy, but cleared later, and the observations were carried out with almost complete success. With the astrographic telescope Mr. Davidson secured 15 out of 18 photographs showing the required stellar images. Totality lasted 6 minutes, and the average exposure of the plates was 5 to 6 seconds. Dr. Crommelin with the other lens had 7 successful plates out of 8. The unsuccessful plates were spoiled for this purpose by the clouds, but show the remarkable prominence very well.

When the plates were developed the astrographic images were found to be out of focus. This is attributed to the effect of the sun's heat on the coelostat mirror. The images were fuzzy and quite different from those on the check-plates secured at night before and after the eclipse. Fortunately the mirror which fed the 4-inch lens was not affected, and the star images secured with this lens were good and similar to those got by the night-plates. The observers stayed on in Brazil until July to secure the field in the night sky at the altitude of the eclipse epoch and under identical instrumental conditions.

The plates were measured at Greenwich immediately after the observers' return. Each plate was measured twice over by Messrs. Davidson and Furner, and I am satisfied that such faults as lie in the results are in the plates themselves and not in the measures. The figures obtained may be briefly summarized as follows: The astrographic plates gave $0''.97$ for the displacement at the limb when the scale-value was determined from the plates themselves, and $1''.40$ when the scale-value was assumed from the check plates. But the much better plates gave for the displacement at the limb $1''.98$, Einstein's predicted value being $1''.75$. Further, for these plates the agreement was all that could be expected....

After a careful study of the plates I am prepared to say that there can be no doubt that they confirm Einstein's prediction. A very definite result has been obtained that light is deflected according to Einstein's law of gravitation.

Professor A. S. Eddington
ROYAL OBSERVATORY

Mr. Cottingham and I left the other observers at Madeira and arrived at *Principe* on April 23.... We soon realized that the prospect of a clear sky at the end of May was not very good. Not even a heavy thunderstorm on the morning of the eclipse, three weeks after the end of the wet season, saved the situation. The sky was completely cloudy at first contact, but about half an hour before totality we began to see glimpses of the sun's crescent through the clouds. We carried out our program exactly as arranged, and the sky must have been clearer towards the end of totality. Of the 16 plates taken during the five minutes of totality the first ten showed no stars at all; of the later plates two showed five stars each, from which a result could be obtained. Comparing them with the check-plates secured at Oxford before we went out, we obtained as the final result from the two plates for the value of the displacement of the limb $1''.6 \pm 0.3$.... This result supports the figures obtained at Sobral....

I will pass now to a few words on the meaning of the result. It points to the larger of the two possible values of the deflection. The simplest interpretation of the bending of the ray is to consider it as an effect of the weight of light. We know that momentum is carried along on the path of a beam of light. Gravity in acting creates momentum in a direction different from that of the path of the ray and so causes it to bend. For the half-effect we have to assume that gravity obeys Newton's law; for the full effect which has been obtained we must assume that gravity obeys the new law proposed by Einstein. This is one of the most crucial tests between Newton's law and the proposed new law. Einstein's law had already indicated a perturbation, causing the orbit of Mercury to revolve. That confirms it for relatively small velocities. Going to the limit, where the speed is that of light, the perturbation is increased in such a way as to double the curvature of the path, and this is now confirmed.

This effect may be taken as proving Einstein's *law* rather than his *theory*. It is not affected by the failure to detect the displacement of Fraunhofer lines on the sun. If this latter failure is confirmed it will not affect Einstein's law of gravitation, but it will affect the views on which the law was arrived at. The law is right, though the fundamental ideas underlying it may yet be questioned....

One further point must be touched upon. Are we to attribute the displacement to the gravitational field and not to the refracting matter around the sun? The refractive index required to produce the result at a distance of 15' from the sun would be that given by gases at a pressure of 1/60 to 1/200 of an atmosphere. This is of too great a density considering the depth through which the light would have to pass.

Sir J. J. Thomson
President of the Royal Society

... If the results obtained had been only that light was affected by gravitation, it would have been of the greatest importance. Newton, did, in fact, suggest this very point in his "Optics," and his suggestion would presumably have led to the half-value. But this result is not an isolated one; it is part of a whole continent of scientific ideas affecting the most fundamental concepts of physics.... This is the most important result obtained in connection with the theory of gravitation since Newton's day, and it is fitting that it should be announced at a meeting of the society so closely connected with him.

The difference between the laws of gravitation of Einstein and Newton come only in special cases. The real interest of Einstein's theory lies not so much in his results as in the method by which he gets them. If his theory is right, it makes us take an entirely new view of gravitation. If it is sustained that Einstein's reasoning holds good—and it has survived two very severe tests in connection with the perihelion of mercury and the present eclipse—then it is the result of one of the highest achievements of human thought. The weak point in the theory is the great difficulty in expressing it. It would seem that no one can understand the new law of gravitation without a thorough knowledge of the theory of invariants and of the calculus of variations.

One other point of physical interest arises from the discussion. Light is deflected in passing near huge bodies of matter. This involves alterations in the electric and magnetic field. This, again, implies the existence of electric and magnetic forces outside matter—forces at present unknown, though some idea of their nature may be got from the results of this expedition.

1

From a report in *The Observatory*, of the Joint Eclipse Meeting of the Royal Society and the Royal Astronomical Society, November 6, 1919. ↑

BY THE SAME AUTHOR

EMINENT CHEMISTS OF OUR TIME

A non-technical account of the more remarkable achievements in the realm of chemistry as exemplified by the life and work of the more modern chemists. There is hardly a chemist of note whose work is not mentioned in connection with one or another of the eleven following: Perkin and Coal Tar Dyes; Mendeléeff and the Periodic Law; Richards and Atomic Weights; Ramsay and the Gases of the Atmosphere; Van 't Hoff and Physical Chemistry; Arrhenius and The Theory of Electrolytic Dissociation; Moissan and the Electric Furnace; Curie and Radium; Victor Meyer and the Rise of Organic Chemistry; Remsen and the Rise of Chemistry in America; Fischer and the Chemistry of Foods.

250 Pages 5 × 7½ Illustrated Cloth $2.50

BOOKS OF INTEREST

THE NATURE OF MATTER AND ELECTRICITY

AN OUTLINE OF MODERN VIEWS

By DANIEL F. COMSTOCK, S.B., Ph.D.

Engineer and Associate Professor of Theoretical Physics In the Massachusetts Institute of Technology, and

LEONARD T. TROLAND, S.B., A.M., Ph.D.

Instructor In Harvard University

225 Pages 5½ × 8 Cloth 22 Illustrations 11 Plates Postpaid $2.50

THE MYSTERY OF MATTER AND ENERGY

RECENT PROGRESS AS TO THE STRUCTURE OF MATTER

By A. C. CREHORE, Ph.D.

172 Pages 4½ × 6½ Cloth 8 Plates and Folding Charts Postpaid $1.00

THE ATOM

By A. C. CREHORE, Ph.D.

250 Pages 5 × 7½ 6 Illustrations $2.00

CPSIA information can be obtained
at www.ICGtesting.com
Printed in the USA
LVHW031526171121
703566LV00019B/723/J